SKELETAL STRUCTURES

Matrix methods of linear structural analysis using influence coefficients

by

C. M. BOMMER

B.Sc. (Eng.), D.I.C., C. Eng., M.I.C.E., A.M. Inst. H.E.

and

D. A. SYMONDS

B.Sc., D.I.C.

CROSBY LOCKWOOD & SON, LTD

26 OLD BROMPTON ROAD, LONDON, S.W. 7.

First published 1968
Printed in Hungary by
Egyetemi Printing House

SKELETAL STRUCTURES

Matrix methods

PREFACE

Most problems in structural analysis require the solution of simultaneous equations. In the past, these equations have been solved either by untidy methods of iteration or methods involving over-simplified assumptions. These were often suitable for only one type of problem.

Such methods were indispensable to the designer, but it meant that students were taught different methods for different problems. This made it harder for the student to obtain a true understanding of structural behaviour, and, therefore, limited his scope.

The object of this book is to introduce a method of analysing all types of linear skeletal structures.

The text is in two parts. The first section gives an introduction to matrix algebra. This is a technique which simplifies the handling and solution of simultaneous equations.

The second section covers the classification of structures, indeterminacy and the flexibility and stiffness methods of analysis. Simple examples are included to clarify the main points.

A summary showing the steps involved in each method of analysis is given to help the reader to put the theory into practice. The relative merits of each approach are discussed with regard to the structure under consideration and the means of computation available.

The effects of temperature, lack of fit, settlement of supports and the calculation of deflections are also covered.

Computer programmes are now readily available for the analysis of a variety of structures and many designers would be less reluctant to use these facilities if they possessed a certain understanding of the way the results are obtained.

In the text that follows simple problems have been used to illustrate the essence of the subject. Matrix methods, when applied to such trivial problems, will seem unduly lengthy as their full advantage is only apparent when applied to complex problems.

ACKNOWLEDGEMENTS

The writing of this book was inspired by the D.I.C. lectures and those given by J. C. de C. Henderson at the Imperial College of Science and Technology, London.

CONTENTS

Part I

MATRIX ALGEBRA

INTRODUCTION

Matrix algebra provides a neat system of handling large numbers of simultaneous equations and linear transformations. Matrices were developed by A. Cayley in 1857 but they have only recently been introduced into engineering. They now have an important role in structural analysis and, indeed, in any field in which large numbers of equations have to be handled.

A matrix may be defined as a rectangular array of numbers or algebraic symbols in rows and columns. This array may be represented by a single symbol. A matrix is denoted by enclosing the array in square or rounded brackets. Unlike a determinant, usually represented by two vertical lines either side of the array, it cannot be evaluated.

Only that part of matrix theory necessary for an understanding of Part II is given here. For those-wishing to study matrices in more detail, a list of references is given on page 106.

The main object is to enable the reader to solve the equations arising from the analysis of structures. The method is given in an orderly manner so that each stage in the calculations may be checked. A desk calculator may be used for up to about eight equations, but for a greater number electronic computers are normally used.

DEFINITIONS AND NOTATIONS

Consider, for example, the set of linear equations

$$\begin{aligned} 3x+2y+4z&=1 \\ x-3y-5z&=2 \\ 7x+y+5z&=0 \end{aligned}$$

which in matrix form can be written

$$\begin{bmatrix} 3 & 2 & 4 \\ 1 & -3 & -5 \\ 7 & 1 & 5 \end{bmatrix} \begin{bmatrix} x \\ y \\ z \end{bmatrix} = \begin{bmatrix} 1 \\ 2 \\ 0 \end{bmatrix}$$

$$(3\times 3) \quad (3\times 1)\ (3\times 1)$$

In a more general form, consider the following set of linear equations

$$\begin{aligned} a_{11}x_1 + a_{12}x_2 + a_{13}x_3 + \ldots + a_{1n}x_n &= c_1 \\ a_{21}x_1 + a_{22}x_2 + a_{23}x_3 + \ldots + a_{2n}x_n &= c_2 \\ a_{31}x_1 + a_{32}x_2 + a_{33}x_3 + \ldots + a_{3n}x_n &= c_3 \\ \vdots \qquad\qquad & \quad \vdots \\ a_{m1}x_1 + a_{m2}x_2 + a_{m3}x_3 + \ldots + a_{mn}x_n &= c_m \end{aligned}$$

which again can be represented thus

$$\begin{bmatrix} a_{11} & a_{12} & a_{13} & \cdots & a_{1n} \\ a_{21} & a_{22} & a_{23} & \cdots & a_{2n} \\ a_{31} & a_{32} & a_{33} & \cdots & a_{3n} \\ \cdot & \cdot & \cdot & \cdots & \cdot \\ a_{m1} & a_{m2} & a_{m3} & \cdots & a_{mn} \end{bmatrix} \begin{bmatrix} x_1 \\ x_2 \\ x_3 \\ \cdot \\ x_n \end{bmatrix} = \begin{bmatrix} c_1 \\ c_2 \\ c_3 \\ \cdot \\ c_m \end{bmatrix}$$

$$(m\times n) \qquad (n\times 1)\ (m\times 1)$$

which may be written as

$$\boldsymbol{A} \cdot \boldsymbol{x} = \boldsymbol{c}$$

$$(m\times n)\quad (n\times 1)\quad (m\times 1)$$

The set of linear equations have thus been separated into three separate matrices $\boldsymbol{A}$, $\boldsymbol{x}$ and $\boldsymbol{c}$ containing the coefficients a_{11} to a_{mn}, the variables we wish to evaluate x_1 to x_n, and the constants c_1 to c_m respectively.

The coefficients a_{11} to a_{mn} are called the ELEMENTS of the matrix $\boldsymbol{A}$.

The horizontal lines of coefficients are called the ROWS and the vertical lines of coefficients are called the COLUMNS of the matrix.

A matrix with m rows and n columns is said to be of the ORDER $(m\times n)$ and is called an "m by n matrix". Matrices containing more than one row and column are denoted by upper-case heavy type letters $\boldsymbol{A}$, $\boldsymbol{B}$, etc., or by $[a_{rs}]$, $[b_{rs}]$ etc., i.e., the general element enclosed in a square or round brackets.

In the double-subscript notation for the elements, the first subscript refers to the row, and the second refers to the column containing the given element.

A matrix $\boldsymbol{A}$ of order $(m \times n)$ can be written

$$\underset{(m\times n)}{\boldsymbol{A}} \equiv [a_{rs}] \equiv \begin{bmatrix} a_{11} & a_{12} & a_{13} \dots a_{1n} \\ a_{21} & a_{22} & a_{23} \dots a_{2n} \\ \cdot & \cdot & \cdot \quad \cdots \cdot \\ a_{m1} & a_{m2} & a_{m3} \dots a_{mn} \end{bmatrix}$$

If $m=n$ then the matrix is called a SQUARE MATRIX.
For example

$$\underset{(3\times 3)}{\boldsymbol{A}} = \begin{bmatrix} a_{11} & a_{12} & a_{13} \\ a_{21} & a_{22} & a_{23} \\ a_{31} & a_{32} & a_{33} \end{bmatrix}$$

If $n=1$ then the order of the matrix is $(m \times 1)$ and it is called a COLUMN MATRIX or COLUMN VECTOR.
For example

$$\begin{bmatrix} a_{11} \\ a_{21} \\ \vdots \\ a_{m1} \end{bmatrix} \equiv \begin{bmatrix} a_1 \\ a_2 \\ \vdots \\ a_m \end{bmatrix} \equiv \{a_1 \quad a_2 \dots a_m\} \equiv \boldsymbol{a}$$

If $m=1$ then the order of the matrix is $(1 \times n)$ and it is called a ROW MATRIX or ROW VECTOR
For example

$$[a_{11} \quad a_{12} \dots a_{1n}] \equiv [a_1 \quad a_2 \dots a_n] \equiv \boldsymbol{a}^{\mathrm{T}} \quad (\text{or } \boldsymbol{a}')$$

Row and column matrices are represented by a lower-case heavy type letter.

If $\boldsymbol{A}=[a_{rs}]$, then the matrix whose typical element is $[a_{sr}]$ is said to be the TRANSPOSE of $\boldsymbol{A}$, and denoted by $\boldsymbol{A}^{\mathrm{T}}$ or $\boldsymbol{A}'$ i.e, the rows of $\boldsymbol{A}$ become the columns of $\boldsymbol{A}^{\mathrm{T}}$ or vice versa.
For example

$$\text{If} \quad \underset{(2\times 3)}{\boldsymbol{A} = \begin{bmatrix} a_{11} & a_{12} & a_{13} \\ a_{21} & a_{22} & a_{23} \end{bmatrix}} \quad \text{then} \quad \underset{(3\times 2)}{\boldsymbol{A}^{\mathrm{T}} = \begin{bmatrix} a_{11} & a_{21} \\ a_{12} & a_{22} \\ a_{13} & a_{23} \end{bmatrix}}$$

Consider a (4×4) square matrix

$$\begin{bmatrix} a_{11} & a_{12} & a_{13} & a_{14} \\ a_{21} & a_{22} & a_{23} & a_{24} \\ a_{31} & a_{32} & a_{33} & a_{34} \\ a_{41} & a_{42} & a_{43} & a_{44} \end{bmatrix}$$

The elements a_{11}, a_{22}, a_{33}, and a_{44} lie on the PRINCIPAL DIAGONAL
The elements a_{12}, a_{23}, and a_{34} lie on the SUPER DIAGONAL
The elements a_{21}, a_{32} and a_{43} lie on the SUB DIAGONAL

If the transpose of a square matrix is the same as the matrix itself, i.e., $\boldsymbol{A}=\boldsymbol{A}^{\mathrm{T}}$, then the matrix is SYMMETRIC
For example

$$\begin{bmatrix} 4 & -3 & 1 \\ -3 & 6 & 2 \\ 1 & 2 & 8 \end{bmatrix}$$
(3×3)

i.e. symmetric about the principal diagonal.

If a square matrix $\boldsymbol{A}=-\boldsymbol{A}^{\mathrm{T}}$, then it must be ANTISYMMETRIC and its principal diagonal must be zero.
For example

$$\begin{bmatrix} 0 & -2 & 1 \\ 2 & 0 & -3 \\ -1 & 3 & 0 \end{bmatrix}$$
(3×3)

A square matrix whose elements are all zero except those on the principal diagonal is named a DIAGONAL MATRIX

A UNIT MATRIX is a diagonal matrix whose pricipal diagonal elements are all unity and is denoted by $\boldsymbol{I}$.
For example

$$\boldsymbol{I}=\text{diag. }\{1 \quad 1 \quad 1\}\equiv\begin{bmatrix} 1 & 0 & 0 \\ 0 & 1 & 0 \\ 0 & 0 & 1 \end{bmatrix}$$
(3×3)

A NULL or ZERO MATRIX is one of any order in which all the elements are zero.
For example

$$\begin{bmatrix} 0 & 0 \\ 0 & 0 \\ 0 & 0 \end{bmatrix}=\boldsymbol{0}$$
(3×2)

A square matrix whose elements on the upper side of the principal diagonal are all zero is called a LOWER TRIANGULAR MATRIX
For example

$$\begin{bmatrix} l_{11} & 0 & 0 & 0 \\ l_{21} & l_{22} & 0 & 0 \\ l_{31} & l_{32} & l_{33} & 0 \\ l_{41} & l_{42} & l_{43} & l_{44} \end{bmatrix}=\boldsymbol{L}$$
(4×4)

A square matrix whose elements on the lower side of the principal diagonal are all zero is called an UPPER TRIANGULAR MATRIX.

For example

$$\underset{(4\times4)}{\begin{bmatrix} u_{11} & u_{12} & u_{13} & u_{14} \\ 0 & u_{22} & u_{23} & u_{24} \\ 0 & 0 & u_{33} & u_{34} \\ 0 & 0 & 0 & u_{44} \end{bmatrix}} = \boldsymbol{U}$$

PROPERTIES OF MATRICES

Matrix addition and subtraction

Two matrices may be added or subtracted if, and only if, they are of the same order. By adding or subtracting the corresponding elements of the two matrices, the resulting elements form a third matrix which is of the same order.

For example

$$\begin{array}{ccccc} \boldsymbol{A} & + & \boldsymbol{B} & = & \boldsymbol{C} \\ \begin{bmatrix} a_{11} & a_{12} & a_{13} \\ a_{21} & a_{22} & a_{23} \end{bmatrix} & & \begin{bmatrix} b_{11} & b_{12} & b_{13} \\ b_{21} & b_{22} & b_{23} \end{bmatrix} & & \begin{bmatrix} c_{11} & c_{12} & c_{13} \\ c_{21} & c_{22} & c_{23} \end{bmatrix} \\ (2\times3) & & (2\times3) & & (2\times3) \end{array}$$

where $a_{rs}+b_{rs}=c_{rs}$

Matrices of the same order are CONFORMABLE for addition and subtraction. Matrix addition or subtraction is COMMUTATIVE and ASSOCIATIVE.

i.e. $\boldsymbol{A}+\boldsymbol{B}=\boldsymbol{B}+\boldsymbol{A}$

and $\boldsymbol{A}+(\boldsymbol{B}+\boldsymbol{C})=(\boldsymbol{A}+\boldsymbol{B})+\boldsymbol{C}$

Matrix multiplication

Scalar Multiplication

It follows from the above rule for addition, that by multiplying a matrix by a single element (scalar), the elements of the resulting matrix are each multiplied by the scalar.

For example

Consider the scalar $\lambda=3$

and the matrix $\boldsymbol{A}=\underset{(2\times3)}{\begin{bmatrix} a_{11} & a_{12} & a_{13} \\ a_{21} & a_{22} & a_{23} \end{bmatrix}}$

Then

$$\lambda \cdot \boldsymbol{A} = 3 \cdot \boldsymbol{A} = \boldsymbol{A} + \boldsymbol{A} + \boldsymbol{A} = \begin{bmatrix} 3a_{11} & 3a_{12} & 3a_{13} \\ 3a_{21} & 3a_{22} & 3a_{23} \end{bmatrix}$$

Matrix Multiplication

By definition, the process of multiplication is rows into columns. Thus, if row $\boldsymbol{a}^{\mathrm{T}} = [a_{11} \quad a_{12} \quad a_{13} \ \dots \ a_{1n}]$
and column

$$\boldsymbol{b} = \begin{bmatrix} b_{11} \\ b_{21} \\ \vdots \\ b_{n1} \end{bmatrix}$$

then

$$\underset{(1 \times n)}{\boldsymbol{a}^{\mathrm{T}} \cdot \boldsymbol{b} = [a_{11} \quad a_{12} \quad a_{13} \ \dots \ a_{1n}]} \ \underset{(n \times 1)}{\begin{bmatrix} b_{11} \\ b_{21} \\ b_{31} \\ \vdots \\ b_{n1} \end{bmatrix}}$$

$$= [a_{11}b_{11} + a_{12}b_{21} + a_{13}b_{31} + \dots + a_{1n}b_{n1}]$$

$$= \underset{(1 \times 1)}{[c_{11}]}$$

$$\underset{(1 \times n)}{\boldsymbol{a}^{\mathrm{T}}} \cdot \underset{(n \times 1)}{\boldsymbol{b}} = \underset{(1 \times 1)}{\boldsymbol{c}} = \underset{(1 \times n)}{\boldsymbol{b}^{\mathrm{T}}} \cdot \underset{(n \times 1)}{\boldsymbol{a}}$$

The product of a row into a column is only possible if the number of elements in each is the same. The resulting quantity is a single element or scalar i.e., a matrix of order (1×1).

Let us now consider the product of two matrices $\boldsymbol{A} \cdot \boldsymbol{B} = \boldsymbol{C}$.

The process of multiplication is rows into columns i.e., the rth row of $\boldsymbol{A}$ multiplied into the sth column of $\boldsymbol{B}$ yields the element of $\boldsymbol{C}$ in the rth row and s^{th} column i.e. c_{rs}.

Clearly the above product is only possible if the number of columns of $\boldsymbol{A}$ equals the number of rows of $\boldsymbol{B}$ i.e. they are CONFORMABLE IN MULTIPLICATION. The resultant matrix $\boldsymbol{C}$ possesses the same number of rows as $\boldsymbol{A}$ and the same number of columns as $\boldsymbol{B}$.

i.e.

$$\underset{(m \times p)}{\boldsymbol{A}} \cdot \underset{(p \times n)}{\boldsymbol{B}} = \underset{(m \times n)}{\boldsymbol{C}}$$

For example

$$\underset{(4\times 2)}{A} \cdot \underset{(2\times 3)}{B} = \underset{(4\times 3)}{C}$$

$$\underset{(4\times 2)}{\begin{bmatrix} a_{11} & a_{12} \\ a_{21} & a_{22} \\ a_{31} & a_{32} \\ a_{41} & a_{42} \end{bmatrix}} \underset{(2\times 3)}{\begin{bmatrix} b_{11} & b_{12} & b_{13} \\ b_{21} & b_{22} & b_{23} \end{bmatrix}} \quad \underset{(4\times 3)}{\begin{bmatrix} (a_{11}b_{11}+a_{12}b_{21}) & (a_{11}b_{12}+a_{12}b_{22}) & (a_{11}b_{13}+a_{12}b_{23}) \\ (a_{21}b_{11}+a_{22}b_{21}) & (a_{21}b_{12}+a_{22}b_{22}) & (a_{21}b_{13}+a_{22}b_{23}) \\ (a_{31}b_{11}+a_{32}b_{21}) & (a_{31}b_{12}+a_{32}b_{22}) & (a_{31}b_{13}+a_{32}b_{23}) \\ (a_{41}b_{11}+a_{42}b_{21}) & (a_{41}b_{12}+a_{42}b_{22}) & (a_{41}b_{13}+a_{42}b_{23}) \end{bmatrix}}$$

$$= \underset{(4\times 3)}{\begin{bmatrix} c_{11} & c_{12} & c_{13} \\ c_{21} & c_{22} & c_{23} \\ c_{31} & c_{32} & c_{33} \\ c_{41} & c_{42} & c_{43} \end{bmatrix}}$$

but $\underset{(2\times 3)}{B} \cdot \underset{(4\times 2)}{A}$ is not possible generally.

i.e. Matrix multiplication is said to be NON-COMMUTATIVE

In general $A \cdot B \neq B \cdot A$

For example

$$A = \underset{(2\times 3)}{\begin{bmatrix} 3 & 2 & 1 \\ 4 & 1 & 5 \end{bmatrix}} \qquad B = \underset{(3\times 2)}{\begin{bmatrix} 1 & 0 \\ 3 & 1 \\ 2 & 3 \end{bmatrix}}$$

$$A \cdot B = AB$$

$$\underset{(2\times 3)}{\begin{bmatrix} 3 & 2 & 1 \\ 4 & 1 & 5 \end{bmatrix}} \underset{(3\times 2)}{\begin{bmatrix} 1 & 0 \\ 3 & 1 \\ 2 & 3 \end{bmatrix}} = \underset{(2\times 2)}{\begin{bmatrix} 11 & 5 \\ 17 & 16 \end{bmatrix}}$$

$$B \cdot A = BA$$

$$\underset{(3\times 2)}{\begin{bmatrix} 1 & 0 \\ 3 & 1 \\ 2 & 3 \end{bmatrix}} \underset{(2\times 3)}{\begin{bmatrix} 3 & 2 & 1 \\ 4 & 1 & 5 \end{bmatrix}} \underset{(3\times 3)}{\begin{bmatrix} 3 & 2 & 1 \\ 13 & 7 & 8 \\ 18 & 7 & 17 \end{bmatrix}}$$

In $B \cdot A$, A is PRE-MULTIPLIED by B

In $A \cdot B$, A is POST-MULTIPLIED by B

Consider the conformable product of three matrices $\boldsymbol{A}$, $\boldsymbol{B}$ and $\boldsymbol{C}$

$$\underset{(m\times n)}{\boldsymbol{A}} \cdot \underset{(n\times p)}{\boldsymbol{B}} \cdot \underset{(p\times q)}{\boldsymbol{C}} \equiv \underset{(m\times p)}{(\boldsymbol{AB})} \cdot \underset{(p\times q)}{\boldsymbol{C}}$$

$$\equiv \underset{(m\times n)}{\boldsymbol{A}} \cdot \underset{(n\times q)}{(\boldsymbol{BC})}$$

i.e. Matrix multiplication is said to be ASSOCIATIVE.

Reversal Rules

$$(\boldsymbol{A}\cdot\boldsymbol{B})^{\mathrm{T}}=\boldsymbol{B}^{\mathrm{T}}\cdot\boldsymbol{A}^{\mathrm{T}}$$
$$\neq\boldsymbol{A}^{\mathrm{T}}\cdot\boldsymbol{B}^{\mathrm{T}} \quad \text{in general}$$
$$(\boldsymbol{A}\cdot\boldsymbol{B})^{-1}=\boldsymbol{B}^{-1}\cdot\boldsymbol{A}^{-1}$$
$$\neq\boldsymbol{A}^{-1}\cdot\boldsymbol{B}^{-1} \quad \text{in general}$$

For Example

$$\boldsymbol{A} \cdot \boldsymbol{B} = \boldsymbol{AB}$$

$$\begin{bmatrix}2 & 3\\1 & 2\end{bmatrix}\begin{bmatrix}-1 & 2\\-1 & 3\end{bmatrix} \quad \begin{bmatrix}-5 & 13\\-3 & 8\end{bmatrix}$$

$$\therefore \quad (\boldsymbol{AB})^{\mathrm{T}}=\begin{bmatrix}-5 & -3\\13 & 8\end{bmatrix}$$

$$\boldsymbol{B}^{\mathrm{T}} \cdot \boldsymbol{A}^{\mathrm{T}} = \boldsymbol{B}^{\mathrm{T}}\boldsymbol{A}^{\mathrm{T}}$$

$$\begin{bmatrix}-1 & -1\\2 & 3\end{bmatrix}\begin{bmatrix}2 & 1\\3 & 2\end{bmatrix} \quad \begin{bmatrix}-5 & -3\\13 & 8\end{bmatrix}$$

SOME ELEMENTARY MATRIX OPERATIONS

(1) Multiplication of a row (or column) by a number.
(2) The interchange of two rows (or columns).
(3) Addition of a multiple of one row (or column) to another row (or column).

In general, to operate on the columns of a matrix, post-multiply by the matrix $\boldsymbol{E}$. Similarly, to operate on the rows of a matrix, pre-multiply by the matrix $\boldsymbol{E}$. The matrix $\boldsymbol{E}$ is a square matrix made up from a unit matrix, but with one or any of its elements replaced by the appropriate number.

Example (a) Construct a matrix $\boldsymbol{E}$ operating on a matrix $\boldsymbol{A}$ which will multiply row 1 by 4

$$\underset{(3\times 3)}{\begin{bmatrix}4&0&0\\0&1&0\\0&0&1\end{bmatrix}} \cdot \underset{(3\times 4)}{\begin{bmatrix}1&2&3&4\\5&6&7&8\\9&10&11&12\end{bmatrix}} = \underset{(3\times 4)}{\begin{bmatrix}4&8&12&16\\5&6&7&8\\9&10&11&12\end{bmatrix}}$$

$$\boldsymbol{E} \cdot \boldsymbol{A} = \boldsymbol{EA}$$

(b) Construct a matrix $\boldsymbol{E}$ which, operating on $\boldsymbol{A}$, will multiply column 3 by 2.

$$\boldsymbol{A} \cdot \boldsymbol{E} = \boldsymbol{AE}$$

$$\underset{(3\times 4)}{\begin{bmatrix}1&2&3&4\\5&6&7&8\\9&10&11&12\end{bmatrix}} \underset{(4\times 4)}{\begin{bmatrix}1&0&0&0\\0&1&0&0\\0&0&2&0\\0&0&0&1\end{bmatrix}} \underset{(3\times 4)}{\begin{bmatrix}1&2&6&4\\5&6&14&8\\9&10&22&12\end{bmatrix}}$$

(c) Construct a matrix $\boldsymbol{E}$ operating on $\boldsymbol{A}$ which will reverse row 1 and row 3 of $\boldsymbol{A}$.

$$\boldsymbol{E} \cdot \boldsymbol{A} = \boldsymbol{EA}$$

$$\underset{(3\times 3)}{\begin{bmatrix}0&0&1\\0&1&0\\1&0&0\end{bmatrix}} \underset{(3\times 4)}{\begin{bmatrix}1&2&3&4\\5&6&7&8\\9&10&11&12\end{bmatrix}} \underset{(3\times 4)}{\begin{bmatrix}9&10&11&12\\5&6&7&8\\1&2&3&4\end{bmatrix}}$$

(d) Construct a matrix $\boldsymbol{E}$ which, operating on $\boldsymbol{A}$, will subtract $2\times$Column 2 from Column 4.

$$\boldsymbol{A} \cdot \boldsymbol{E} = \boldsymbol{AE}$$

$$\underset{(3\times 4)}{\begin{bmatrix}1&2&3&4\\5&6&7&8\\9&10&11&12\end{bmatrix}} \underset{(4\times 4)}{\begin{bmatrix}1&0&0&0\\0&1&0&-2\\0&0&1&0\\0&0&0&1\end{bmatrix}} \underset{(3\times 4)}{\begin{bmatrix}1&2&3&0\\5&6&7&-4\\9&10&11&-8\end{bmatrix}}$$

It can be seen that it is possible to operate on a given matrix $\boldsymbol{A}$ by a square matrix $\boldsymbol{E}$ which will alter the elements of the matrix without altering its order. The application of this will become more apparent in the section describing matrix inversion.

SOME SPECIAL MATRICES

By definition

$\boldsymbol{e} =$ Col. (1, 1...1) of order $(n \times 1)$ as appropriate in the context.
and $\boldsymbol{e}_k =$ Col. (0, 0...1...0) of order $(n \times 1)$ where all elements are zero except the kth element which is unity.

Pre-multiplying by $\boldsymbol{e}$, i.e., $\boldsymbol{e}^{\mathrm{T}} \cdot \boldsymbol{A}$ produces a row matrix whose elements are the sums of the columns of $\boldsymbol{A}$.

For example

$$\underset{(1\times 3)}{\begin{bmatrix}1 & 1 & 1\end{bmatrix}} \cdot \underset{(3\times 3)}{\begin{bmatrix}1 & 2 & 3\\ 4 & 5 & 6\\ 7 & 8 & 9\end{bmatrix}} = \underset{(1\times 3)}{\begin{bmatrix}12 & 15 & 18\end{bmatrix}}$$

$$\boldsymbol{e}^{\mathrm{T}} \cdot \boldsymbol{A} = \boldsymbol{e}^{\mathrm{T}}\boldsymbol{A}$$

Post-multiplying by $\boldsymbol{e}$, i.e. $\boldsymbol{A} \cdot \boldsymbol{e}$ produces a column matrix whose elements are the sums of the rows of $\boldsymbol{A}$.

For example

$$\boldsymbol{A} \cdot \boldsymbol{e} = \boldsymbol{A}\boldsymbol{e}$$

$$\underset{(3\times 3)}{\begin{bmatrix}1 & 2 & 3\\ 4 & 5 & 6\\ 7 & 8 & 9\end{bmatrix}} \underset{(3\times 1)}{\begin{bmatrix}1\\ 1\\ 1\end{bmatrix}} \quad \underset{(3\times 1)}{\begin{bmatrix}6\\ 15\\ 24\end{bmatrix}}$$

The product $\boldsymbol{e}^{\mathrm{T}} \cdot \boldsymbol{A} \cdot \boldsymbol{e}$ adds all the elements of $\boldsymbol{A}$ giving a single element matrix.

Pre-multiplying by $\boldsymbol{e}_k$ i.e., $\boldsymbol{e}_k^{\mathrm{T}} \cdot \boldsymbol{A}$ picks out the kth row of $\boldsymbol{A}$.

For example

$$\boldsymbol{e}_2^{\mathrm{T}} \cdot \boldsymbol{A} = \boldsymbol{e}_2^{\mathrm{T}}\boldsymbol{A}$$

$$\underset{(1\times 3)}{\begin{bmatrix}0 & 1 & 0\end{bmatrix}} \underset{(3\times 3)}{\begin{bmatrix}1 & 2 & 3\\ 4 & 5 & 6\\ 7 & 8 & 9\end{bmatrix}} \quad \underset{(1\times 3)}{\begin{bmatrix}4 & 5 & 6\end{bmatrix}}$$

Similarly, post-multiplying by $\boldsymbol{e}_k$ i.e., $\boldsymbol{A} \cdot \boldsymbol{e}_k$, picks out the kth column of $\boldsymbol{A}$.

For example

$$\boldsymbol{A} \cdot \boldsymbol{e}_2 = \boldsymbol{A}\boldsymbol{e}$$

$$\underset{(3\times 3)}{\begin{bmatrix}1 & 2 & 3\\ 4 & 5 & 6\\ 7 & 8 & 9\end{bmatrix}} \underset{(3\times 1)}{\begin{bmatrix}0\\ 1\\ 0\end{bmatrix}} \quad \underset{(3\times 1)}{\begin{bmatrix}2\\ 5\\ 8\end{bmatrix}}$$

The application of the above technique is in the step-by-step checks which are made in various matrix operations shown in the following example, and in the section on matrix inversion.

Example (e) Form $\boldsymbol{A}\cdot\boldsymbol{B}$ with suitable checks.

$$\underset{(3\times5)}{\boldsymbol{A}} \quad\cdot\quad \underset{(5\times3)}{\boldsymbol{B}} \quad=\quad \underset{(3\times3)}{\boldsymbol{C}}$$

$$\begin{bmatrix}3&8&7&15&9\\-3&2&1&4&-2\\4&-1&8&7&3\end{bmatrix}\begin{bmatrix}1&2&3\\4&5&-6\\3&-2&1\\1&3&-2\\5&4&3\end{bmatrix}\qquad\begin{bmatrix}116&113&-35\\2&6&-34\\46&20&21\end{bmatrix}$$

$$\boldsymbol{e}^{\mathrm{T}}\cdot\boldsymbol{A}=(4\quad 9\quad 16\quad 26\quad 10)\qquad\qquad \boldsymbol{e}^{\mathrm{T}}\cdot\boldsymbol{C}=(\underset{\checkmark}{164}\quad \underset{\checkmark}{139}\quad \underset{\checkmark}{-48})$$

$$\text{i.e.}\quad (\boldsymbol{e}^{\mathrm{T}}\cdot\boldsymbol{A})\cdot\boldsymbol{B}=\boldsymbol{e}^{\mathrm{T}}\cdot\boldsymbol{C}$$

MATRIX INVERSION

Matrix inversion is a process in matrix algebra similar to division in ordinary algebra. The analysis of a structure having α degrees of indeterminacy involves the solution of α linearly independent equations. In matrix form these may be written $\boldsymbol{A}\cdot\boldsymbol{x}=\boldsymbol{c}$ where $\boldsymbol{A}$ is a non-singular symmetrical square matrix of order $(\alpha\times\alpha)$, and $\boldsymbol{x}$ and $\boldsymbol{c}$ are column matrices of order $(\alpha\times1)$. The matrix $\boldsymbol{A}$ is said to be singular if the determinant formed from the elements of $\boldsymbol{A}$ is equal to zero.

Given a square matrix $\boldsymbol{A}$, it is possible to find another matrix $\boldsymbol{B}$ of the same order such that $\boldsymbol{A}\cdot\boldsymbol{B}=\boldsymbol{I}$ provided the above conditions are satisfied. If such a matrix $\boldsymbol{B}$ exists, it is called the INVERSE of $\boldsymbol{A}$ and is written as $\boldsymbol{A}^{-1}$.

If $\boldsymbol{A}\cdot\boldsymbol{x}=\boldsymbol{c}$, then by pre-multiplying both sides of the above by $\boldsymbol{A}^{-1}$, we have $\boldsymbol{A}^{-1}\cdot\boldsymbol{A}\cdot\boldsymbol{x}=\boldsymbol{A}^{-1}\boldsymbol{c}$ and since $\boldsymbol{A}^{-1}\cdot\boldsymbol{A}=\boldsymbol{A}\cdot\boldsymbol{A}^{-1}=\boldsymbol{I}$, we get

$$\boldsymbol{I}\cdot\boldsymbol{x}=\boldsymbol{A}^{-1}\cdot\boldsymbol{c}$$

$$\text{or}\qquad \boldsymbol{x}=\boldsymbol{A}^{-1}\cdot\boldsymbol{c}$$

Having evaluated $\boldsymbol{A}^{-1}$, the values of the elements of $\boldsymbol{x}$ can be obtained by matrix multiplication.

The inverse of a square, non-singular matrix can be found in a variety of ways, and four methods are described below.

(1) Inverse of a Symmetric Matrix by Factorisation

If $\boldsymbol{A}$ is a symmetric matrix having real elements, then there exists a lower triangular matrix $\boldsymbol{L}$ such that $\boldsymbol{A}=\boldsymbol{L}\cdot\boldsymbol{L}^{\mathrm{T}}$.

Pre-multiplying by L^{-1}, we have $\quad L^{-1}\cdot A = L^{-1}\cdot L\cdot L^{\mathrm{T}}$

or $\quad L^{-1}\cdot A = L^{\mathrm{T}}$

and post-multiplying by A^{-1} $\quad L^{-1}\cdot A\cdot A^{-1} = L^{\mathrm{T}}\cdot A^{-1}$

$$L^{-1} = L^{\mathrm{T}}\cdot A^{-1}$$

To simplify the computation, it is convenient at this stage to modify the multiplication operations to (column)×(column) and (row)×(row).

$(R\times C)$ — signifies conventional matrix multiplication
$(C\times C)$ — column×column multiplication
$(R\times R)$ — row×row multiplication

re-writing the above equations in the new form

Conventional Form	**Modified Form**
$A = L\cdot L^{\mathrm{T}}$	$A = L^{\mathrm{T}}\cdot L^{\mathrm{T}}$
$(R\times C)$	$(C\times C)$
$L^{-1} = L^{\mathrm{T}}\cdot A^{-1}$	$L^{-1} = L^{\mathrm{T}}\cdot (A^{-1})^{\mathrm{T}}$
$(R\times C)$	$(R\times R)$

The setting out of the matrix A for inversion is indicated below. Numbers in brackets indicate the order of operations. Consider a (3×3) matrix A.

$A\cdot e = S_1$ **Explanation of Checks**

$$A = \begin{bmatrix} a_{11} & a_{12} & a_{13} \\ a_{21} & a_{22} & a_{23} \\ a_{31} & a_{32} & a_{33} \end{bmatrix} \qquad \begin{matrix} (1) \\ (2) \\ (3) \end{matrix}$$

$(C\times C)$

$A = L\cdot L^{\mathrm{T}}$ $(R\times C)$
$S_1 = A\cdot e = L\cdot L^{\mathrm{T}}\cdot e$
or $S_1 = (L^{\mathrm{T}})S_2$ $(C\times C)$

$L^{\mathrm{T}}\cdot e = S_2$

$$L^{\mathrm{T}} = \begin{bmatrix} (4) & (5) & (6) \\ 0 & (8) & (9) \\ 0 & 0 & (11) \end{bmatrix} \qquad \begin{matrix} (7) \\ (10) \\ (12) \end{matrix}$$

$(R\times R)$

$$\text{diag. } L^{-1} = \frac{1}{l_{rr}} = \{(13) \quad (14) \quad (15)\}$$

S_1 (Rewrite)

$$(A^{-1})^{\mathrm{T}} = \begin{bmatrix} (21) & (20a) & (18a) \\ (20) & (19) & (17a) \\ (18) & (17) & (16) \end{bmatrix}$$

(1′) $\mathrm{Col}_r(A^{-1})^{\mathrm{T}}\times S_1 = 1.00000000$
(2′)
(3′)

This method is known as Fox's variation of Cholesky's Method. A complete example illustrating this method is given in Part II. A step-by-step description is given below to clarify the operations.

$$A=\begin{bmatrix} a_{11} & a_{12} & a_{13} \\ a_{21} & a_{22} & a_{23} \\ a_{31} & a_{32} & a_{33} \end{bmatrix} \qquad \begin{matrix} S_1 \\ (1) \\ (2) \\ (3) \end{matrix}$$

$$L^{\mathrm{T}}=\begin{bmatrix} l_{11} & l_{21} & l_{31} \\ 0 & l_{22} & l_{32} \\ 0 & 0 & l_{33} \end{bmatrix} \qquad \begin{matrix} S_2\ (C\times C) \\ (7) \\ (10) \\ (12) \end{matrix}$$

$$\text{diag. } L^{-1}=\{l^{11} \quad l^{22} \quad l^{33}\} \qquad S_1\ (R\times R)$$

$$(A^{-1})^{\mathrm{T}}=\begin{bmatrix} a^{11} & a^{21} & a^{31} \\ a^{12} & a^{22} & a^{32} \\ a^{13} & a^{23} & a^{33} \end{bmatrix} \qquad \begin{matrix} (1') \\ (2') \\ (3') \end{matrix}$$

To find checks S_1

$(1)=a_{11}+a_{12}+a_{13}$
$(2)=a_{21}+a_{22}+a_{23}$
$(3)=a_{31}+a_{32}+a_{33}$

To find L^{T}: Column $\times$ Column Multiplication: $C_r(L^{\mathrm{T}})\times C_s(L^{\mathrm{T}})=a_{rs}$

(4) (Col. 1 of L^{T})(Col. 1 of L^{T})$=a_{11}$ i.e. $l_{11}\cdot l_{11}=a_{11}$ Hence $l_{11}=\sqrt{a_{11}}$

(5) (Col. 1 of L^{T})(Col. 2 of L^{T})$=a_{12}$ i.e. $l_{11}\cdot l_{21}=a_{12}$ Hence $l_{21}=\dfrac{a_{12}}{l_{11}}$

(6) (Col. 1 of L^{T})(Col. 3 of L^{T})$=a_{13}$ i.e. $l_{11}\cdot l_{31}=a_{13}$ Hence $l_{31}=\dfrac{a_{13}}{l_{11}}$

(7) $=l_{11}+l_{21}+l_{31}=$ 1st element of S_2
Hence apply 1st check $l_{11}\cdot(7)$ and if this is equal to (1), tick (7) and proceed with computation.

(8) (Col. 2 of L^{T})(Col. 2 of L^{T})$=a_{22}$ i.e. $l_{21}^2+l_{22}^2=a_{22}$ Hence l_{22}

(9) (Col. 2 of L^{T})(Col. 3 of L^{T})$=a_{23}$ i.e. $l_{21}\cdot l_{31}+l_{22}\cdot l_{32}=a_{23}$ Hence l_{32}

(10) $=0+l_{22}+l_{32}=$ 2nd element of S_2. Hence apply 2nd check $l_{21}\cdot(7)+l_{22}\cdot(10)$, and if this is equal to (2), tick (10) and proceed with computation.

(11) (Col. 3 of L^{T})(Col. 3 of L^{T})$=a_{33}$ i.e. $l_{31}^2+l_{32}^2+l_{33}^2=a_{33}$ Hence l_{33}

(12) $=0+0+l_{33}=$ 3rd element of S_2. Hence apply 3rd check $l_{31}\cdot(7)+l_{32}\cdot(10)+l_{33}\cdot(12)$ and if this is equal to (3) tick (12) and proceed with computation.

To find diagonal elements of $\boldsymbol{L}^{-1}$

(13) $l^{11}=\dfrac{1}{l_{11}}$

(14) $l^{22}=\dfrac{1}{l_{22}}$

(15) $l^{33}=\dfrac{1}{l_{33}}$

To find elements of $(\boldsymbol{A}^{-1})^{\mathrm{T}}$: Row$\times$Row Multiplication:

$\text{Row}_r(\boldsymbol{L}^{\mathrm{T}})\times\text{Row}_s(\boldsymbol{A}^{-1})^{\mathrm{T}}=l^{rs}$

Rewrite elements of S_1 i.e., (1′), (2′) and (3′)

(16) (Row 3 of $\boldsymbol{L}^{\mathrm{T}}$)(Row 3 of $(\boldsymbol{A}^{-1})^{\mathrm{T}}$)$=l^{33}$

i.e. $l_{33}\cdot a^{33}=l^{33}$ Hence a^{33}

(17) (Row 2 of $\boldsymbol{L}^{\mathrm{T}}$)(Row 3 of $(\boldsymbol{A}^{-1})^{\mathrm{T}}$)$=0$

i.e. $l_{22}\cdot a^{23}+l_{32}\cdot a^{33}=0$ Hence a^{23}

(18) (Row 1 of $\boldsymbol{L}^{\mathrm{T}}$)(Row 3 of $(\boldsymbol{A}^{-1})^{\mathrm{T}}$)$=0$

i.e. $l_{11}\cdot a^{13}+l_{21}\cdot a^{23}+l_{31}\cdot a^{33}=0$ Hence a^{13}

By symmetry (17a)$=$(17) and (18a)$=$(18)

Apply check (Col. 3 of $(\boldsymbol{A}^{-1})^{\mathrm{T}}$)(Col. S_1) should give unity. If correct, proceed with computation

(19) (Row 2 of $\boldsymbol{L}^{\mathrm{T}}$)(Row 2 of $(\boldsymbol{A}^{-1})^{\mathrm{T}}$)$=l^{22}$

i.e. $l_{22}\cdot a^{22}+l_{32}\cdot a^{32}=l_{22}$ Hence a^{22}

(20) (Row 1 of $\boldsymbol{L}^{\mathrm{T}}$)(Row 2 of $(\boldsymbol{A}^{-1})^{\mathrm{T}}$)$=0$ Hence a^{12}

By symmetry (20a)$=$(20)

Apply check (Col. 2 of $(\boldsymbol{A}^{-1})^{\mathrm{T}}$)(Col. S_1) should give unity. If correct, proceed with computation.

(21) (Row 1 of $\boldsymbol{L}^{\mathrm{T}}$)(Row 1 of $(\boldsymbol{A}^{-1})^{\mathrm{T}}$)$=l^{11}$ Hence a^{11}

Final check, (Col. 1 of $(\boldsymbol{A}^{-1})^{\mathrm{T}}$)(Col. S_1) should give unity.

Since the matrix is symmetrical, $(\boldsymbol{A}^{-1})^{\mathrm{T}}=\boldsymbol{A}^{-1}$

A symmetric matrix also has a symmetric inverse.

This method is the most suitable for solution with a desk calculating machine for up to about 8×8 order matrices.

It should be noted that the inverse of a diagonal matrix is also a diagonal matrix the elements of which are the reciprocals of the corresponding elements of $\boldsymbol{A}$.

For example

$$\begin{bmatrix} a_{11} & 0 & 0 \\ 0 & a_{22} & 0 \\ 0 & 0 & a_{33} \end{bmatrix}\begin{bmatrix} \frac{1}{a_{11}} & 0 & 0 \\ 0 & \frac{1}{a_{22}} & 0 \\ 0 & 0 & \frac{1}{a_{33}} \end{bmatrix}=\begin{bmatrix} 1 & 0 & 0 \\ 0 & 1 & 0 \\ 0 & 0 & 1 \end{bmatrix}$$

$$\text{i.e.}\qquad \boldsymbol{A}\quad\cdot\quad\boldsymbol{A}^{-1}\quad=\quad\boldsymbol{I}$$

Furthermore, the inverse of an Upper or Lower triangular matrix is also an Upper or Lower triangular matrix, the principal diagonal elements of which are the reciprocals of the original matrix. This can be easily verified.

$$\boldsymbol{L} \quad \cdot \quad \boldsymbol{L}^{-1} \quad = \quad \boldsymbol{I}$$

$$\begin{bmatrix} l_{11} & 0 & 0 \\ l_{21} & l_{22} & 0 \\ l_{31} & l_{32} & l_{33} \end{bmatrix} \begin{bmatrix} \frac{1}{l_{11}} & 0 & 0 \\ l^{21} & \frac{1}{l_{22}} & 0 \\ l^{31} & l^{32} & \frac{1}{l_{33}} \end{bmatrix} \quad \begin{bmatrix} 1 & 0 & 0 \\ 0 & 1 & 0 \\ 0 & 0 & 1 \end{bmatrix}$$

It follows, therefore, that to carry out the operations $R_r(\boldsymbol{L}^{\mathrm{T}}) \times R_s(\boldsymbol{A}^{-1})^{\mathrm{T}} = l^{rs}$ in the above, the only elements required are those on the principal diagonal of $\boldsymbol{L}^{-1}$,

i.e. $\quad \boldsymbol{L}^{\mathrm{T}} \quad \cdot \quad (\boldsymbol{A}^{-1})^{\mathrm{T}} \quad = \quad \boldsymbol{L}^{-1}$

$$\begin{bmatrix} l_{11} & l_{21} & l_{31} \\ 0 & l_{22} & l_{32} \\ 0 & 0 & l_{33} \end{bmatrix} \begin{bmatrix} a^{11} & a^{21} & a^{31} \\ a^{12} & a^{22} & a^{32} \\ a^{13} & a^{23} & a^{33} \end{bmatrix} \quad \begin{bmatrix} \frac{1}{l_{11}} & 0 & 0 \\ X & \frac{1}{l_{22}} & 0 \\ X & X & \frac{1}{l_{33}} \end{bmatrix}$$

where X are the elements of $\boldsymbol{L}^{-1}$ which are not required to obtain $(\boldsymbol{A}^{-1})^{\mathrm{T}}$.

(2) Inverse of a Matrix by use of its adjugate matrix

Consider a square matrix $\boldsymbol{A}$ of order n. For $\boldsymbol{A}$ to have an inverse, the determinant of the elements of $\boldsymbol{A}$ must not be zero, i.e. $|A| \neq 0$.
The MINOR of an element a_{rs} is the $(n-1)^{\text{th}}$ order determinant formed by removal of the rth row and s^{th} column from A.
The CO-FACTOR of a_{rs} is formed by multiplying the minor of a_{rs} by $(-1)^{(r+s)}$ and is denoted by A_{rs}

The numerical value of $|A| = a_{r1}A_{r1} + a_{r2}A_{r2} + \ldots + a_{rn}A_{rn}$ when expanded using the rth or $|A| = a_{1s}A_{1s} + a_{2s}A_{2s} + \ldots + a_{ns}A_{ns}$ when expanded using the sth column.
The expansion of $|A|$ by false co-factors is zero.
For example. $a_{21} \cdot A_{11} + a_{22} \cdot A_{12} + a_{23} \cdot A_{13} = 0$
The ADJUGATE of any square matrix $\boldsymbol{A}$ denoted adj. $\boldsymbol{A}$ is formed by transposing the matrix whose elements are the co-factors of the elements of $\boldsymbol{A}$, thus
if

$$\boldsymbol{A} = \begin{bmatrix} a_{11} & a_{12} & a_{13} \\ a_{21} & a_{22} & a_{23} \\ a_{31} & a_{32} & a_{33} \end{bmatrix}$$

then

$$\text{adj. } \boldsymbol{A} = \begin{bmatrix} A_{11} & A_{21} & A_{31} \\ A_{12} & A_{22} & A_{32} \\ A_{13} & A_{23} & A_{33} \end{bmatrix}$$

The product of $\boldsymbol{A}$. (adj. $\boldsymbol{A}$) gives

$$\begin{bmatrix} a_{11} & a_{12} & a_{13} \\ a_{21} & a_{22} & a_{23} \\ a_{31} & a_{32} & a_{33} \end{bmatrix} \begin{bmatrix} A_{11} & A_{21} & A_{31} \\ A_{12} & A_{22} & A_{32} \\ A_{13} & A_{23} & A_{33} \end{bmatrix} = \begin{bmatrix} |A| & 0 & 0 \\ 0 & |A| & 0 \\ 0 & 0 & |A| \end{bmatrix}$$

$$\text{i.e. } \boldsymbol{A} \cdot (\text{adj. } \boldsymbol{A}) = |A| \cdot \boldsymbol{I}$$

$$\text{or } \boldsymbol{A} \frac{\text{adj. } \boldsymbol{A}}{|A|} = \boldsymbol{I}$$

$$\therefore \quad \boldsymbol{A}^{-1} = \frac{\text{adj. } \boldsymbol{A}}{|A|}$$

This method of inversion is not suitable for matrices of a higher order than 4×4, since the calculation becomes very tedious. It is not possible to check each stage of the calculation with this method and it is, therefore, not recommended. However, for second order matrices it provides a quick solution.

For example

$$\boldsymbol{A} = \begin{bmatrix} 1 & 2 \\ 3 & 4 \end{bmatrix} \qquad \text{adj. } \boldsymbol{A} = \begin{bmatrix} 4 & -2 \\ -3 & 1 \end{bmatrix}$$

$$|A| = 1 \times 4 - 2 \times 3 = -2$$

$$\therefore \boldsymbol{A}^{-1} = -\frac{1}{2} \begin{bmatrix} 4 & -2 \\ -3 & 1 \end{bmatrix} = \begin{bmatrix} -2 & 1 \\ 1.5 & -0.5 \end{bmatrix}$$

Check

$$S_1 = \boldsymbol{e}^{\mathrm{T}} \cdot \boldsymbol{A}$$

$$\boldsymbol{e}^{\mathrm{T}} \cdot \boldsymbol{A} \cdot \boldsymbol{A}^{-1} = \boldsymbol{e}^{\mathrm{T}} \cdot \boldsymbol{I}$$

$$= \boldsymbol{e}^{\mathrm{T}}$$

$$\begin{array}{cc} \boldsymbol{A} & \boldsymbol{A}^{-1} \\ \begin{bmatrix} 1 & 2 \\ 3 & 4 \end{bmatrix} & \begin{bmatrix} -2 & 1 \\ 1.5 & -0.5 \end{bmatrix} \end{array}$$

$$S_1 = \boldsymbol{e}^{\mathrm{T}} \cdot \boldsymbol{A} = (4 \quad 6) \quad (1.0 \quad 1.0) = \boldsymbol{e}^{\mathrm{T}}$$

(3) **Inverse of a Matrix by Partitioning**

The time taken to invert a matrix is roughly proportional to n^3 where n is the order of the matrix. The inversion of matrices of higher order is more conveniently done by partitioning the matrix into smaller matrices called sub-matrices. These are isolated by dotted lines as shown below. The matrix, whose elements are sub-matrices, conforms to the usual laws of matrix algebra.

Let

$$A = \left[\begin{array}{cc:cc} a_{11} & a_{12} & a_{13} & a_{14} \\ a_{21} & a_{22} & a_{23} & a_{24} \\ \hdashline a_{31} & a_{32} & a_{33} & a_{34} \\ a_{41} & a_{42} & a_{43} & a_{44} \end{array}\right]$$

$$\equiv \left[\begin{array}{c:c} \boldsymbol{B} & \boldsymbol{C} \\ \hdashline \boldsymbol{D} & \boldsymbol{E} \end{array}\right] \quad \text{in partitioned form.}$$

For the purpose of inversion, it is necessary that sub-matrices $\boldsymbol{B}$ and $\boldsymbol{E}$ are both square and non-singular.

Let

$$\boldsymbol{A}^{-1} = \begin{bmatrix} \boldsymbol{W} & \boldsymbol{X} \\ \boldsymbol{Y} & \boldsymbol{Z} \end{bmatrix} \quad \text{in partitioned form.}$$

It follows, therefore, that

$$\boldsymbol{A} \cdot \boldsymbol{A}^{-1} = \boldsymbol{I}$$

i.e.

$$\begin{bmatrix} \boldsymbol{B} & \boldsymbol{C} \\ \boldsymbol{D} & \boldsymbol{E} \end{bmatrix} \begin{bmatrix} \boldsymbol{W} & \boldsymbol{X} \\ \boldsymbol{Y} & \boldsymbol{Z} \end{bmatrix} = \begin{bmatrix} \boldsymbol{I} & \boldsymbol{0} \\ \boldsymbol{0} & \boldsymbol{I} \end{bmatrix}$$

Carrying out the multiplication of the $\boldsymbol{A} \cdot \boldsymbol{A}^{-1}$, we have

$$\boldsymbol{B} \cdot \boldsymbol{W} + \boldsymbol{C} \cdot \boldsymbol{Y} = \boldsymbol{I}$$

$$\boldsymbol{B} \cdot \boldsymbol{X} + \boldsymbol{C} \cdot \boldsymbol{Z} = \boldsymbol{0}$$

$$\boldsymbol{D} \cdot \boldsymbol{W} + \boldsymbol{E} \cdot \boldsymbol{Y} = \boldsymbol{0}$$

$$\boldsymbol{D} \cdot \boldsymbol{X} + \boldsymbol{E} \cdot \boldsymbol{Z} = \boldsymbol{I}$$

Expressing the unknowns $\boldsymbol{W}$, $\boldsymbol{X}$, $\boldsymbol{Y}$ and $\boldsymbol{Z}$ in terms of the known submatrices $\boldsymbol{A}$, $\boldsymbol{B}$, $\boldsymbol{C}$ and $\boldsymbol{D}$, it can be shown that

$$\boldsymbol{Z} = (\boldsymbol{E} - \boldsymbol{D} \cdot \boldsymbol{B}^{-1} \cdot \boldsymbol{C})^{-1}$$

$$\boldsymbol{Y} = -(\boldsymbol{E} - \boldsymbol{D} \cdot \boldsymbol{B}^{-1} \cdot \boldsymbol{C})^{-1} \cdot \boldsymbol{D} \cdot \boldsymbol{B}^{-1} = -\boldsymbol{Z} \cdot \boldsymbol{D} \cdot \boldsymbol{B}^{-1}$$

$$\boldsymbol{X} = -\boldsymbol{B}^{-1} \cdot \boldsymbol{C} \cdot (\boldsymbol{E} - \boldsymbol{D} \cdot \boldsymbol{B}^{-1} \cdot \boldsymbol{C})^{-1} = -\boldsymbol{B}^{-1} \cdot \boldsymbol{C} \cdot \boldsymbol{Z}$$

$$\boldsymbol{W} = \boldsymbol{B}^{-1} + \boldsymbol{B}^{-1} \cdot \boldsymbol{C}(\boldsymbol{E} - \boldsymbol{D} \cdot \boldsymbol{B}^{-1} \cdot \boldsymbol{C})^{-1} \cdot \boldsymbol{D} \cdot \boldsymbol{B}^{-1} = \boldsymbol{B}^{-1} - \boldsymbol{X} \cdot \boldsymbol{D} \cdot \boldsymbol{B}^{-1}$$

It can be seen that the inversion is reduced to two matrices of a smaller order i.e. $\boldsymbol{B}^{-1}$ and $(\boldsymbol{E} - \boldsymbol{D} \cdot \boldsymbol{B}^{-1} \cdot \boldsymbol{C})^{-1}$. The associated multiplications, additions and subtractions can be carried out more rapidly than inversion of higher order matrices and, therefore, results in considerable time saving.

For a large matrix $\boldsymbol{A}$ of order $n \times n$, the process can be applied by starting with the top left-hand corner of $\boldsymbol{A}$ and adding a row and column at a time until the inverse A^{-1} is obtained. This is known as the Escalator Method.

Each submatrix inversion involves the calculation of Z, Y, X and W from $B^{-1}\cdot C$, $D\cdot B^{-1}$, $D\cdot B^{-1}\cdot C$, and $X\cdot D\cdot B^{-1}$.

Example Compute the inverse of

$$P=\begin{bmatrix}1 & 2 & 3 & 4\\ 2 & 1 & 1 & 5\\ 4 & 1 & 2 & 0\\ 3 & 4 & 1 & 2\end{bmatrix}$$

1*st Operation.* Take the first (2×2) top L.H. side Sub-Matrix

$$\left[\begin{array}{c|c}1 & 2\\ \hline 2 & 1\end{array}\right]$$

$$B^{-1}=\frac{1}{1}=1 \qquad Z=(1-4)^{-1}=-\frac{1}{3}$$

$$B^{-1}\cdot C=1\cdot 2=2 \qquad Y=-\left(-\frac{1}{3}\right)2=\frac{2}{3}$$

$$D\cdot B^{-1}=2\cdot 1=2 \qquad X=-2\left(-\frac{1}{3}\right)=\frac{2}{3}$$

$$D\cdot B^{-1}\cdot C=2\cdot 1\cdot 2=4 \qquad W=1-\left(\frac{2}{3}\right)2=-\frac{1}{3}$$

2*nd Operation.*

$$\left[\begin{array}{cc|c}1 & 2 & 3\\ 2 & 1 & 1\\ \hline 4 & 1 & 2\end{array}\right]$$

From above

$$B^{-1}=\begin{bmatrix}-\frac{1}{3} & \frac{2}{3}\\ \frac{2}{3} & -\frac{1}{3}\end{bmatrix}$$

$$B^{-1}\cdot C=\begin{bmatrix}-\frac{1}{3} & \frac{2}{3}\\ \frac{2}{3} & -\frac{1}{3}\end{bmatrix}\begin{bmatrix}3\\ 1\end{bmatrix}=\begin{bmatrix}-\frac{1}{3}\\ \frac{5}{3}\end{bmatrix} \qquad Z=\left[2\ \ -\frac{1}{3}\right]^{-1}=\frac{3}{5}$$

$$D\cdot B^{-1}=[4\ \ 1]\begin{bmatrix}-\frac{1}{3} & \frac{2}{3}\\ \frac{2}{3} & -\frac{1}{3}\end{bmatrix}=\left[-\frac{2}{3}\ \ \frac{7}{3}\right] \qquad Y=-\frac{3}{5}\left[-\frac{2}{3}\ \ \frac{7}{3}\right]=\left[\frac{2}{5}\ \ -\frac{7}{5}\right]$$

$$D\cdot B^{-1}\cdot C=\left[-\frac{2}{3}\ \ \frac{7}{3}\right]\begin{bmatrix}3\\ 1\end{bmatrix}=\frac{1}{3} \qquad X=-\begin{bmatrix}-\frac{1}{3}\\ \frac{5}{3}\end{bmatrix}\frac{3}{5}=\begin{bmatrix}\frac{1}{5}\\ -1\end{bmatrix}$$

$$X\cdot D\cdot B^{-1}=\begin{bmatrix}\frac{1}{5}\\ -1\end{bmatrix}\left[-\frac{2}{3}\ \ \frac{7}{3}\right]=\begin{bmatrix}-\frac{2}{15} & \frac{7}{15}\\ \frac{2}{3} & -\frac{7}{3}\end{bmatrix}$$

$$W=\begin{bmatrix}-\frac{1}{3} & \frac{2}{3}\\ \frac{2}{3} & -\frac{1}{3}\end{bmatrix}-\begin{bmatrix}-\frac{2}{15} & \frac{7}{15}\\ \frac{2}{3} & -\frac{7}{3}\end{bmatrix}=\begin{bmatrix}-\frac{1}{5} & \frac{1}{5}\\ 0 & 2\end{bmatrix}$$

3rd Operation.

$$\left[\begin{array}{ccc|c} 1 & 2 & 3 & 4 \\ 2 & 1 & 1 & 5 \\ 4 & 1 & 2 & 0 \\ \hline 3 & 4 & 1 & 2 \end{array}\right]$$

From above

$$\boldsymbol{B}^{-1}=\begin{bmatrix} -\frac{1}{5} & \frac{1}{5} & \frac{1}{5} \\ 0 & 2 & -1 \\ \frac{2}{5} & -\frac{7}{5} & \frac{3}{5} \end{bmatrix}$$

$$\boldsymbol{B}^{-1}\cdot\boldsymbol{C}=\begin{bmatrix} -\frac{1}{5} & \frac{1}{5} & \frac{1}{5} \\ 0 & 2 & -1 \\ \frac{2}{5} & -\frac{7}{5} & \frac{3}{5} \end{bmatrix}\begin{bmatrix} 4 \\ 5 \\ 0 \end{bmatrix}=\begin{bmatrix} \frac{1}{5} \\ 10 \\ -\frac{27}{5} \end{bmatrix}$$

$$\boldsymbol{D}\cdot\boldsymbol{B}^{-1}=[3 \quad 4 \quad 1]\begin{bmatrix} -\frac{1}{5} & \frac{1}{5} & \frac{1}{5} \\ 0 & 2 & -1 \\ \frac{2}{5} & -\frac{7}{5} & \frac{3}{5} \end{bmatrix}=\left[-\tfrac{1}{5} \quad \tfrac{36}{5} \quad -\tfrac{14}{5}\right]$$

$$\boldsymbol{D}\cdot\boldsymbol{B}^{-1}\cdot\boldsymbol{C}=\left[-\tfrac{1}{5} \quad \tfrac{36}{5} \quad -\tfrac{14}{5}\right]\begin{bmatrix} 4 \\ 5 \\ 0 \end{bmatrix}=\frac{176}{5}$$

$$\boldsymbol{Z}=\left(2 \; -\frac{176}{5}\right)^{-1}=-\frac{5}{166}$$

$$\boldsymbol{Y}=-\left(-\frac{5}{166}\right)\left[-\tfrac{1}{5} \quad \tfrac{36}{5} \quad -\tfrac{14}{5}\right]=\left[-\tfrac{1}{166} \quad \tfrac{36}{166} \quad -\tfrac{14}{166}\right]$$

$$\boldsymbol{X}=-\begin{bmatrix} \frac{1}{5} \\ 10 \\ -\frac{27}{5} \end{bmatrix}\left(-\frac{5}{166}\right)=\begin{bmatrix} \frac{1}{166} \\ \frac{50}{166} \\ -\frac{27}{166} \end{bmatrix}$$

$$\boldsymbol{X}\cdot\boldsymbol{D}\cdot\boldsymbol{B}^{-1}=\begin{bmatrix} \frac{1}{166} \\ \frac{50}{166} \\ -\frac{27}{166} \end{bmatrix}\left[-\tfrac{1}{5} \quad \tfrac{36}{5} \quad -\tfrac{14}{5}\right]=\begin{bmatrix} -\frac{1}{830} & \frac{36}{830} & -\frac{14}{830} \\ -\frac{10}{166} & \frac{360}{166} & -\frac{140}{166} \\ \frac{27}{830} & -\frac{972}{830} & \frac{378}{830} \end{bmatrix}$$

$$\boldsymbol{W}=\begin{bmatrix} -\frac{1}{5} & \frac{1}{5} & \frac{1}{5} \\ 0 & 2 & -1 \\ \frac{2}{5} & -\frac{7}{5} & \frac{3}{5} \end{bmatrix}-\begin{bmatrix} -\frac{1}{830} & \frac{36}{830} & -\frac{14}{830} \\ -\frac{10}{166} & \frac{360}{166} & -\frac{140}{166} \\ \frac{27}{830} & -\frac{972}{830} & \frac{378}{830} \end{bmatrix}=\begin{bmatrix} -\frac{165}{830} & \frac{130}{830} & \frac{180}{830} \\ \frac{10}{166} & -\frac{28}{166} & -\frac{26}{166} \\ \frac{305}{830} & -\frac{190}{830} & \frac{120}{830} \end{bmatrix}$$

$\therefore$ From above

$$P^{-1}=\begin{bmatrix} -\frac{165}{830} & \frac{130}{830} & \frac{180}{830} & \frac{1}{166} \\ \frac{10}{166} & -\frac{28}{166} & -\frac{26}{166} & \frac{50}{166} \\ \frac{305}{830} & -\frac{190}{830} & \frac{120}{830} & -\frac{27}{166} \\ -\frac{1}{166} & \frac{36}{166} & -\frac{14}{166} & -\frac{5}{166} \end{bmatrix}$$

Check $$e^{\mathrm{T}}\cdot P\cdot P^{-1}=e^{\mathrm{T}}$$

$$\begin{matrix} P & \cdot & P^{-1} \end{matrix}$$

$$\begin{bmatrix} 1 & 2 & 3 & 4 \\ 2 & 1 & 1 & 5 \\ 4 & 1 & 2 & 0 \\ 3 & 4 & 1 & 2 \end{bmatrix}\begin{bmatrix} -\frac{165}{830} & \frac{130}{830} & \frac{180}{830} & \frac{1}{166} \\ \frac{10}{166} & -\frac{28}{166} & -\frac{26}{166} & \frac{50}{166} \\ \frac{305}{830} & -\frac{190}{830} & \frac{120}{830} & -\frac{27}{166} \\ -\frac{1}{166} & \frac{36}{166} & -\frac{14}{166} & -\frac{5}{166} \end{bmatrix}$$

$$e^{\mathrm{T}}\cdot P=(10 \quad 8 \quad 7 \quad 11) \qquad (1 \quad 1 \quad 1 \quad 1)=e^{\mathrm{T}}\cdot P\cdot P^{-1}=e^{\mathrm{T}}$$

Appropriate checks should be applied to all the multiplications carried out as outlined earlier on.

This method may seem tedious, but it must be remembered that the effort required to perform matrix multiplication is relatively simple compared with that required for matrix inversion.

(4) Inverse of a Matrix by Successive Operators

Consider a non-singular matrix A. It is possible to alter the elements of the matrix to a unit matrix by successively operating on matrix A with the E operators described earlier on.

i.e. $$(E_n \ldots E_3\cdot E_2\cdot E_1)\cdot A=I$$
$$E\cdot A=I$$

The matrix product $(E_n\cdot E_{n-1} \ldots E_1)=A^{-1}$

As described earlier, the E matrix is built up successively starting with a unit matrix while A is reduced successively to a unit matrix.

ADJUSTMENT OF AN INVERSE OF A MATRIX RESULTING FROM SMALL ALTERATIONS TO THE PARENT MATRIX

It is sometimes necessary to alter the section properties of a member in a structure for which the matrix has already been inverted. The inverse of the new matrix can be obtained by adjusting the inverse of the original matrix without repeating the complete process of inversion, provided the alterations are not excessive.

Consider a matrix $\boldsymbol{A}$ whose exact inverse $\boldsymbol{A}^{-1}$ has been determined. Suppose that $\boldsymbol{A}$ has been slightly modified to give a matrix $\boldsymbol{X}$ such that $\boldsymbol{X}=\boldsymbol{A}+\delta\boldsymbol{A}$, where $\delta\boldsymbol{A}$ is the matrix representing the difference between $\boldsymbol{X}$ and $\boldsymbol{A}$, and whose elements are small compared with $\boldsymbol{A}$. It is required to compute the inverse of the new matrix $\boldsymbol{X}$.

$$\begin{aligned}\boldsymbol{X}&=\boldsymbol{A}+\delta\boldsymbol{A}\\&=\boldsymbol{A}+\boldsymbol{A}\cdot\boldsymbol{A}^{-1}\cdot\delta\boldsymbol{A}\\&=\boldsymbol{A}\cdot[\boldsymbol{I}+\boldsymbol{A}^{-1}\cdot\delta\boldsymbol{A}]\\ \therefore\ \boldsymbol{X}^{-1}&=[\boldsymbol{A}\cdot[\boldsymbol{I}+\boldsymbol{A}^{-1}\cdot\delta\boldsymbol{A}]]^{-1}\end{aligned}$$

Applying the reversal rule $\boldsymbol{X}^{-1}=[\boldsymbol{I}+\boldsymbol{A}^{-1}\cdot\delta\boldsymbol{A}]^{-1}\cdot\boldsymbol{A}^{-1}$
Expanding the expression by the binomial theorem yields

$$\boldsymbol{X}^{-1}=[\boldsymbol{I}-(\boldsymbol{A}^{-1}\cdot\delta\boldsymbol{A})+(\boldsymbol{A}^{-1}\cdot\delta\boldsymbol{A})^2-(\boldsymbol{A}^{-1}\cdot\delta\boldsymbol{A})^3+\ \ldots]\cdot\boldsymbol{A}^{-1}$$

$\boldsymbol{X}^{-1}$ can, therefore, be obtained from $\boldsymbol{A}^{-1}$ and $\delta\boldsymbol{A}$.

Example

It is required to find the inverse of the matrix

$$\boldsymbol{X}=\begin{bmatrix}4&2&1\\3&7&1\\2&5&5\end{bmatrix}$$

while the inverse of the matrix

$$\boldsymbol{A}=\begin{bmatrix}4&2&1\\3&7&2\\2&5&5\end{bmatrix}$$

is known

$$\boldsymbol{A}^{-1}=\begin{bmatrix}0{\cdot}3165&-0{\cdot}0633&-0{\cdot}0380\\-0{\cdot}1392&0{\cdot}2278&-0{\cdot}0633\\0{\cdot}0126&-0{\cdot}2025&0{\cdot}2785\end{bmatrix}$$

The difference

$$\delta\boldsymbol{A}=\begin{bmatrix}0&0&0\\0&0&-1\\0&0&0\end{bmatrix}$$

The product $\boldsymbol{A}^{-1}\cdot\delta\boldsymbol{A}$ yields

$$\boldsymbol{A}^{-1}\cdot\delta\boldsymbol{A}=\begin{bmatrix}0&0&0{\cdot}0633\\0&0&-0{\cdot}2278\\0&0&0{\cdot}2025\end{bmatrix}$$

Using only the first two terms

$$X^{-1}=[I-(A^{-1}\cdot\delta A)]\,A^{-1}$$

$$=\begin{bmatrix}1 & 0 & -0\cdot0633\\ 0 & 1 & 0\cdot2278\\ 0 & 0 & 0\cdot7975\end{bmatrix}\begin{bmatrix}0\cdot3165 & -0\cdot0633 & -0\cdot0380\\ -0\cdot1392 & 0\cdot2278 & -0\cdot0633\\ 0\cdot0126 & -0\cdot2025 & 0\cdot2785\end{bmatrix}$$

$$=\begin{bmatrix}0\cdot3157 & -0\cdot0505 & -0\cdot0556\\ -0\cdot1363 & 0\cdot1817 & 0\cdot0001\\ 0\cdot0100 & -0\cdot1615 & 0\cdot2221\end{bmatrix}$$

Using the first three terms

$$(A^{-1}\cdot\delta A)^2=\begin{bmatrix}0 & 0 & 0\cdot0128\\ 0 & 0 & -0\cdot0461\\ 0 & 0 & 0\cdot0410\end{bmatrix}$$

$$\therefore\ [I-(A^{-1}\cdot\delta A)+(A^{-1}\cdot\delta A)^2]=\begin{bmatrix}1 & 0 & -0\cdot0505\\ 0 & 1 & 0\cdot1817\\ 0 & 0 & 0\cdot8385\end{bmatrix}$$

$$X^{-1}=\begin{bmatrix}0\cdot3159 & -0\cdot0531 & -0\cdot0521\\ -0\cdot1369 & 0\cdot1910 & -0\cdot0127\\ 0\cdot0106 & -0\cdot1698 & 0\cdot2335\end{bmatrix}$$

The actual value of X^{-1} to 4 places of decimals is given below for comparison.

$$X^{-1}=\begin{bmatrix}0\cdot3158 & -0\cdot0526 & -0\cdot0526\\ -0\cdot1368 & 0\cdot1895 & -0\cdot0105\\ 0\cdot0105 & -0\cdot1684 & 0\cdot2316\end{bmatrix}$$

Part II

STRUCTURAL ANALYSIS

INTRODUCTION

The analysis of a structure involves the determination of stress resultants and displacements at all desired points of the structure under any given loading conditions. A linear, skeletal structure is one whose members can be represented by lines possessing certain rigidity properties. Furthermore, the stress resultants and displacements are directly proportional to the applied loads.

In linear analysis, it is assumed that the deformations due to applied loads are sufficiently small compared with the dimensions of the structure as not to alter its geometry. The principle of superposition also applies. The majority of structures encountered in practice behave in an approximately linear manner.

In solving any structural problem, the following must be considered

(a) The stress-strain relationship of the material of the structure.

(b) The equilibrium of the stress resultants and loads.

(c) The continuity of deformations and displacements.

Statically determinate structures are those which can be analysed by the application of the equations of equilibrium alone. Statically indeterminate, or hyperstatic, structures are those for which the equations of equilibrium are not sufficient by themselves, and the equations of continuity must also be applied.

The methods of analysis of indeterminate structures fall into two basic categories depending on the order in which the equilibrium and continuity conditions are applied.

If the conditions of equilibrium are applied first, the approach is known as the FLEXIBILITY method. This method is also referred to as the LOAD, ACTION OR COMPATIBILITY METHOD.

If the conditions of continuity are used first, the approach is known as the STIFFNESS method, also known as the DISPLACEMENT, DEFORMATION OR EQUILIBRIUM METHOD.

CLASSIFICATIONS, DEFINITIONS AND NOTATIONS

Classification of Skeletal Structures

Skeletal structures may be classified as follows

(i) Direct force structures such as pin-jointed frames which are supported and loaded at the nodes. Only one stress resultant, the axial force (n), may arise.

(ii) Plane frames in which all the members and applied forces lie in the same plane. There are only three possible stress resultants at any point. These are the bending moment and the associated shear force, and the axial force, i.e., (m, s, n).

(iii) Plane frames (grids) in which all the members lie in the same plane, and all the applied loads act normal to the plane of the frame. Again, there are only three possible stress resultants at any point, namely, the normal bending moment, the corresponding shear force and the torque, i.e., (g, q, t).

(iv) Space frames where no limitations are imposed on the geometry or loading, in which a maximum of six stress resultants may occur at any point on the structure, namely, the three mutually perpendicular bending moments and shear forces, i.e. (m, s, n, g, q, t).

Stress Resultants

External forces acting on a structure produce, at any section along a member, certain internal forces and moments which are called STRESS RESULTANTS. The maximum number of stress resultants that can occur at any section is six, the three orthogonal moments and three orthogonal shear forces. These may also be described as the axial force, two moments, two corresponding shear forces and a torque. (Fig. 1)

Each consists of an opposing pair which creates a deformation in an element of length of a member. Stress resultants are denoted by the column matrix $\boldsymbol{x}$.

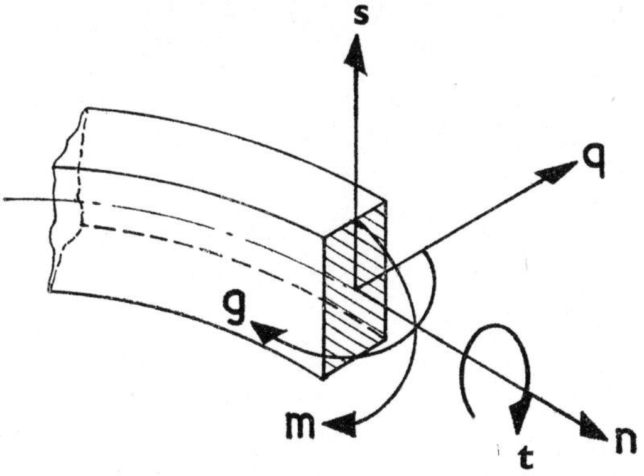

FIG. 1

$$\boldsymbol{x} = \begin{bmatrix} x_1 \\ x_2 \\ x_3 \\ x_4 \\ x_5 \\ x_6 \end{bmatrix} = \begin{bmatrix} m \\ s \\ n \\ g \\ q \\ t \end{bmatrix}$$

Releases

A release is a discontinuity which renders a member incapable of transmitting a stress resultant across that section. There are six releases corresponding to the six stress resultants.

A hinge is an example of a bending moment release about the axis of the hinge which is capable of transmitting all the remaining five stress resultants. A cut releases all six possible stress resultants in a space frame, or all three possible stress resultants in a plane frame.

In general, the six releases $\boldsymbol{r}$ corresponding to the six stress resultants $\boldsymbol{x}$ are denoted as shown below

Stress Resultants *Corresponding Releases* *Notation*

$$\boldsymbol{x} = \begin{bmatrix} m \\ s \\ n \\ g \\ q \\ t \end{bmatrix} = \begin{bmatrix} x_1 \\ x_2 \\ x_3 \\ x_4 \\ x_5 \\ x_6 \end{bmatrix} \qquad \boldsymbol{r} = \begin{bmatrix} r_1 \\ r_2 \\ r_3 \\ r_4 \\ r_5 \\ r_6 \end{bmatrix}$$

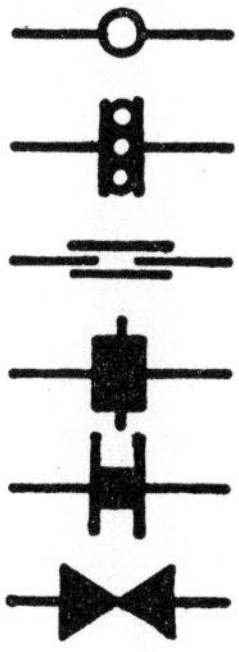

Alternatively, the releases may be represented by zero elements of $\boldsymbol{x}$. For example, a ball-joint releases all three orthogonal moments, and may be represented by

$$\boldsymbol{x} = \begin{bmatrix} 0 \\ x_2 \\ x_3 \\ 0 \\ x_5 \\ 0 \end{bmatrix} = \begin{bmatrix} 0 \\ s \\ n \\ 0 \\ q \\ 0 \end{bmatrix} = \begin{bmatrix} r_1 \\ 0 \\ 0 \\ r_4 \\ 0 \\ r_6 \end{bmatrix}$$

A "ball-joint" is denoted by  and a cut by 

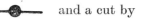

A release does not necessarily occur at a point, but may be continuous along the whole length of a member. A chain, for example, has bending moment releases throughout its length.

A structure may possess certain built-in releases such as the two hinges in a pinned arch, or the two hinges and a roller in a simply supported beam.

In the Flexibility Method, it is necessary to introduce arbitrary releases at convenient points in the structure to make it statically determinate. A completely stiff structure is one which contains no releases.

FLEXIBILITY AND STIFFNESS METHODS

The Flexibility Method

In the Flexibility Method, the structure is first made statically determinate by introducing suitable releases. This results in displacement discontinuities $\boldsymbol{v}_0$ at these releases under the externally applied loads. Suitable pairs of forces and moments called bi-actions $\boldsymbol{p}$ are then applied at the releases in order to restore the continuity of the structure. The magnitude of these bi-actions required to restore continuity in the released structure S_0 is then calculated, usually by an exact method involving the solution of a set of linear simultaneous equations. The effects of the applied loads and the bi-actions acting on the released structure with all external loads removed, are then combined to give the total stress resultants $\boldsymbol{x}_t$ in the structure. The number of simultaneous equations involved is the same as the degree of statical indeterminacy α_s. An example of the Flexibility Method is the Strain-Energy Method.

The Stiffness Method

In the Stiffness Method, on the other hand, the first step is to make the structure kinematically determinate by clamping each of the nodes. This results in stress resultant discontinuities at these nodes under the applied loads. Suitable displacements are then applied at the nodes to restore the equilibrium of the stress resultants. An iterative example of the Stiffness approach is the Hardy Cross Moment Distribution Method. An exact method, which requires the solution of a set of simultaneous equations, is the Slope-Deflection Method.

In the Stiffness Method, the number of simultaneous equations is given by the degree of kinematical indeterminacy α_k.

A Comparison of the Methods

The statical and kinematical indeterminacy gives the number of simultaneous equations to be solved depending on the method adopted. These are give by

$$\alpha_s = P(M - N + 1) - r$$

$$\alpha_k = P(N - 1) + r - c$$

Where P = 3 for plane frames subject to special loadings classified under (ii) and (iii) on page 32.

P = 6 for general loading for space frames classified under (iv) on page 32.

M = Number of members which includes the foundation as a singly-connected system.

N = Number of nodes. The nodes in a structure are located at all points where two or more members meet, at all points of support, and at the free ends.

r = Number of releases.

c = Number of constraints, a constraint being defined as that which prevents any relative degree of freedom between two adjacent nodes connected by a member.

Any part of a structure which is considered to be infinitely stiff (the ground for example) may be replaced by a single node.

The examples shown in Fig. 2 illustrate that the indeterminacy may vary for any given type of structure and that a majority of structures encountered in practice have a lower statical than kinematical degree of indeterminacy.

Structures having a large number of members but only few nodes, normally have a lower kinematical indeterminacy and are, therefore, more suited to analysis by the Stiffness Method.

STATICAL AND KINEMATICAL INDETERMINACY

FOR EXAMPLES (a) TO (h) THE APPLIED LOADS ARE ASSUMED TO LIE IN THE PLANE OF THE STRUCTURE.

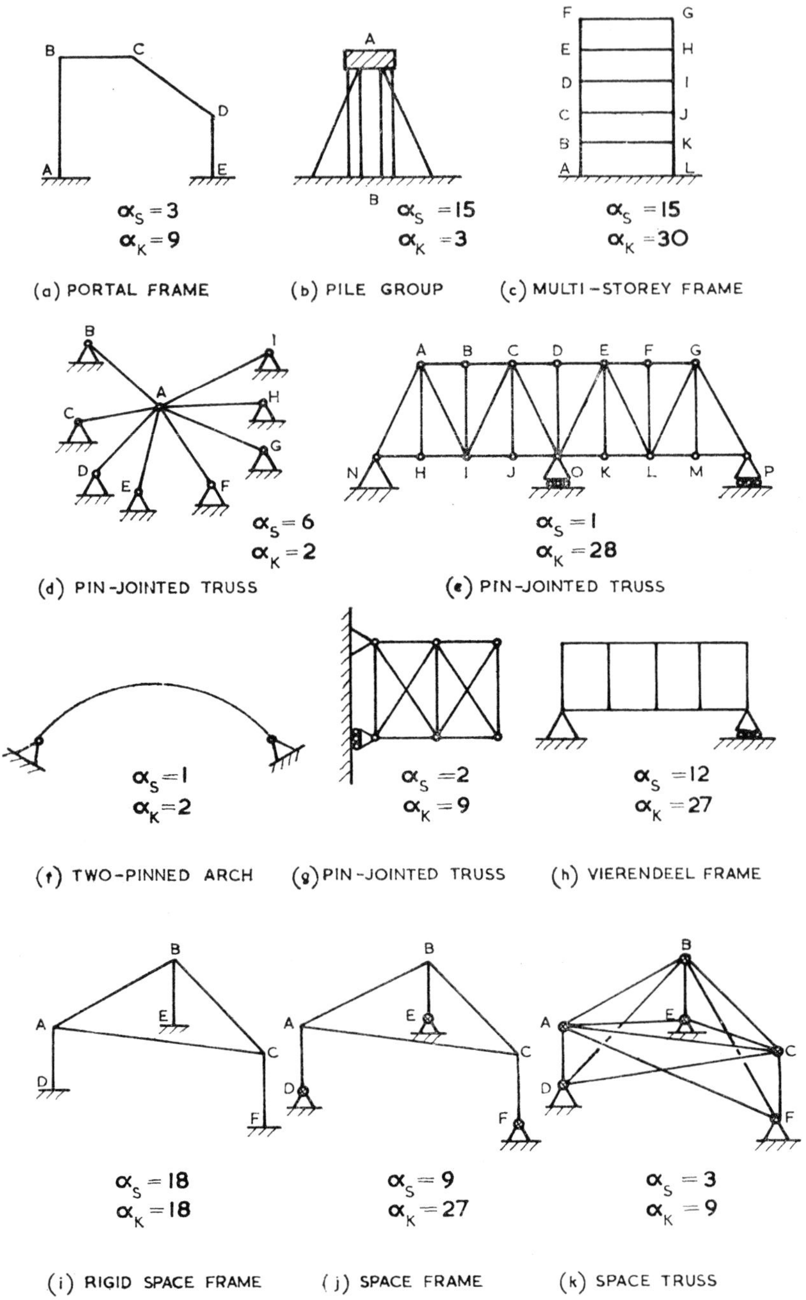

FIG. 2

Most of the work involved in solving a set of simultaneous equations often lies in the matrix inversion. Clearly, if the numerical work is to be carried out with a desk calculating machine, it is best to adopt the method which reduces this work to a minimum. It follows that, for most structures, the Flexibility Method is the most suitable. Where an electronic computer is available, the amount of numerical work is unimportant. The choice of method depends on the simplicity of programming. For this reason, the Stiffness Method is normally used when a computer is employed.

TABLE 1. SUMMARY OF SUITABLE METHODS

Indeterminacy Number		Method to be Used	
α_s	α_k	With Desk Calculator	With Electronic Computer
small	large	Flexibility	Stiffness
large	small	Stiffness	Stiffness
small	small	Flexibility or Stiffness	Stiffness
large	large	Neither	Stiffness

INDETERMINACY

Statical Indeterminacy

The method described is due to J. C. de C. Henderson and W. G. Bickley.

Consider a completely stiff structure S_α shown in Fig. 3. The foundation is also included in the number of members of the complete struc-

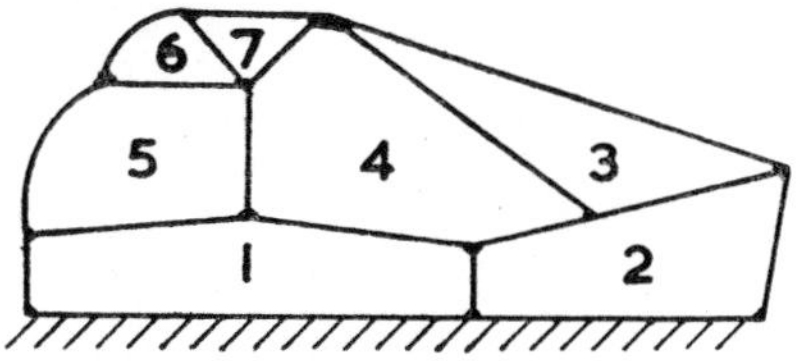

S_α

FIG. 3

ture. The structure is made up of seven loops or rings. The shape of the rings is immaterial. Number of nodes $N=12$, Number of Members $M=18$.

The structure can be made statically determinate by converting it into a 'tree' structure. This is a system in which there is only one path between any two nodes. The property of a 'tree' structure is that the number of nodes is always one more than the number of members i.e., $N_{TREE} = M + 1$. This may be seen from the following examples.

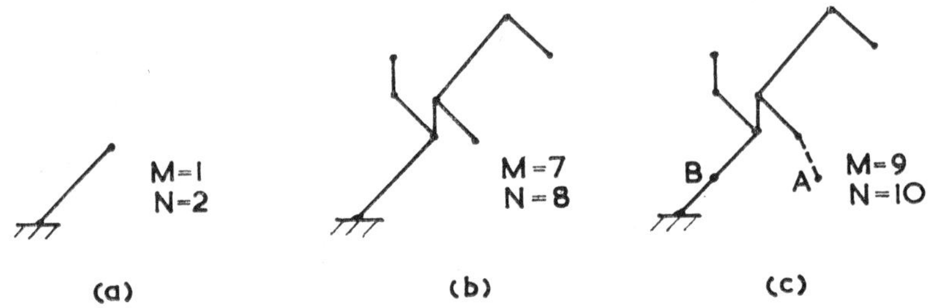

FIG. 4

It can also be seen that the addition of an external member A or an internal node B (Fig. 4 (b) and (c)) merely increases the number of nodes and members by an equal amount, and the relation $N_{TREE} = M + 1$ still holds.

The complete structure can be made into a 'tree' structure by removing seven members, or by disconnecting the members at the nodes so as to break up all the rings. The two alternative forms of the 'tree' structure are shown in Fig. 5.

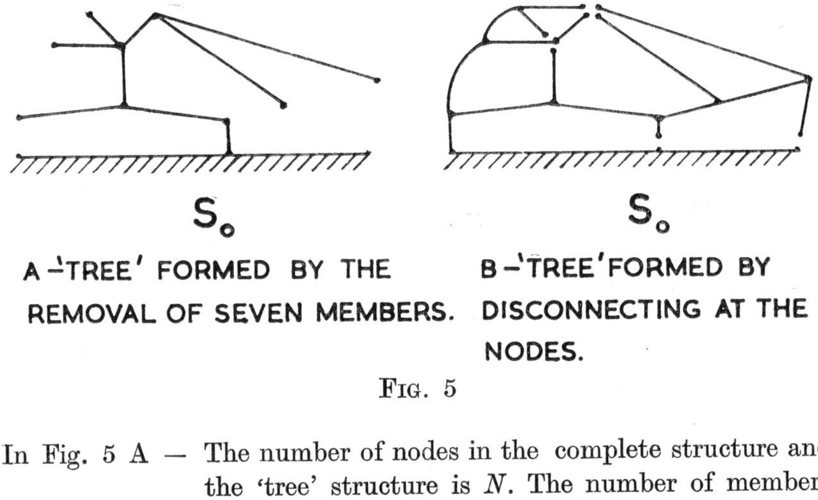

FIG. 5

In Fig. 5 A — The number of nodes in the complete structure and the 'tree' structure is N. The number of members in the 'tree' is $M_{TREE} = N - 1$. The number of members in the complete structure is M. The number of rings R in the structure is given by the expression

$$M - M_{TREE} = M - (N - 1)$$

or

$$R = (M - N + 1)$$

In Fig. 5 B — Each time a ring is broken by disconnecting a member at a node, the number of nodes is increased by one until a 'tree' structure is obtained in which the number of nodes is one more than the number of members. If N is the number of nodes in the complete structure, then the number of rings is given by the expression

$$
\begin{aligned}
R &= N_{\text{TREE}} - N \\
&= (M+1) - N \\
&= (M - N + 1)
\end{aligned}
$$

Now each ring is either 6 or 3 times statically indeterminate depending on the structure and its loading. The degree of statical indeterminacy of a completely stiff structure is, therefore, given by

$$
\begin{aligned}
\alpha_s &= P \cdot R \\
&= P(M - N + 1)
\end{aligned}
$$

where $$P = 6 \text{ or } 3$$

If the actual structure is not completely stiff, but has built-in releases, then the actual degree of statical indeterminacy is given by

$$\underline{\alpha_s = P(M - N + 1) - r}$$

where r is the number of built-in releases.

For a simply supported pin-jointed plane frame, the above expression can be simplified.

Consider an N-girder with completely rigid joints. If two hinges are inserted at each end of every member, the structure would have nodes which could rotate as small mechanisms.

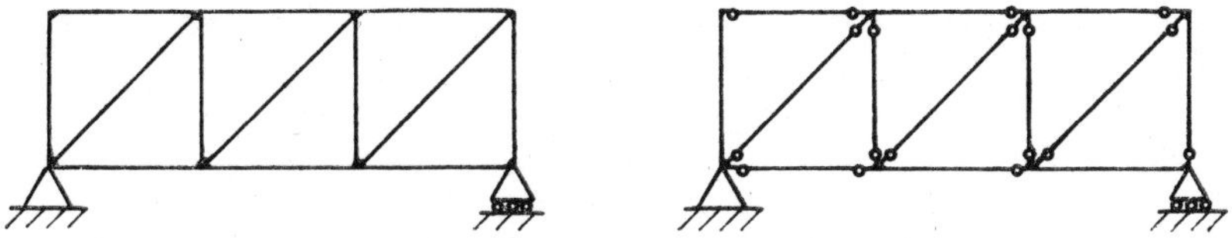

FIG. 6

To prevent this, one hinge is removed from each node. The number of releases in a pin-jointed frame, therefore, is given by the expression

$$r = 2M - N$$

hence $$\alpha_s = 3(M - N + 1) - (2M - N)$$

or $$\underline{\alpha_s = M - 2N + 3}$$

The original rigid N-girder and the corresponding pin-jointed structure are illustrated in Fig. 6.

In a simply supported ball-jointed space frame (Fig. 7), a ball-joint at each end of a member would provide six releases but the members would be free to rotate. The torque is retained in one ball-joint, leaving $5M$ releases. As before, each node can rotate as a small mechanism about three axes, and this requires three constraints to prevent this. Therefore, the total number of releases is given by

$$r = 5M - 3N$$

Hence $$\alpha_s = 6(M - N + 1) - (5M - 3N)$$

or $$\underline{\alpha_s = M - 3N + 6}$$

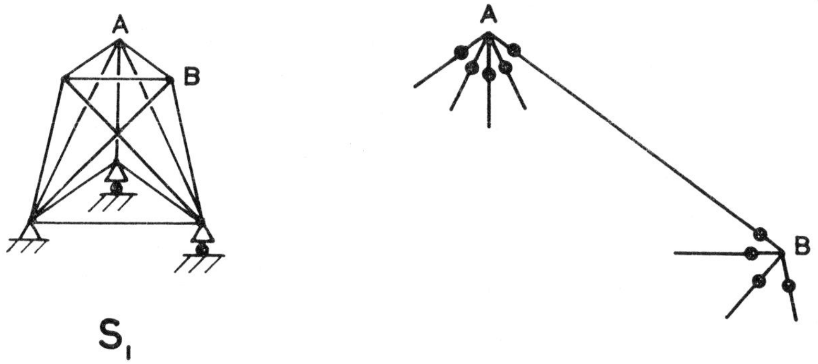

FIG. 7

For examples of statical indeterminacy numbers see Fig. 2. Consider the completely stiff space frame shown in Fig. 2(i). To calculate the degree of statical indeterminacy, it is necessary to include the foundation as a singly-connected system shown by shading in Fig. 8 below.

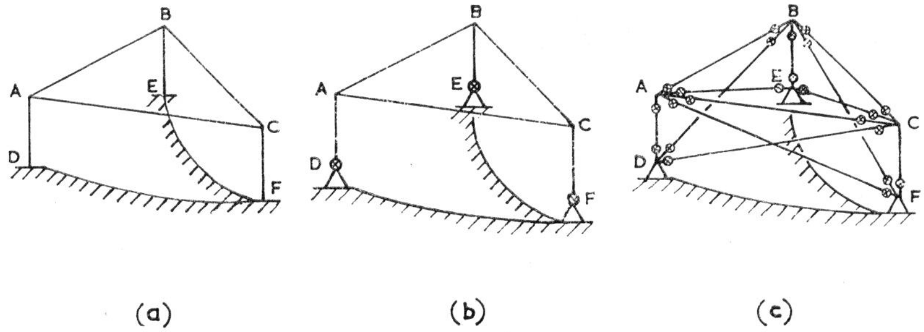

FIG. 8

The degree of statical indeterminacy for the rigid space frame is obtained from the following

$$M=8$$

$$N=6 \qquad \therefore \quad \alpha_s=6(8-6+1)$$

$$r=0 \qquad\qquad =18$$

Now consider the same structure with ball-joints at D, E and F. Each ball-joint releases three stress resultants,

$$\therefore \quad r=9$$

and

$$\alpha_s=18-9$$

$$=9$$

As a further development, consider the ball-jointed space truss of Fig. 2(k). This is repeated in Fig. 8(c) with all the releases shown. $M=14, N=6$ and there are five releases for each of the twelve members excluding the two foundation members DF and EF. As mentioned earlier, it is necessary to provide three moment constraints at each of the nodes A, B and C, making the total number of releases

$$r=12\times5-3\times3=51$$

$$\alpha_s=6(14-6+1)-51$$

$$=3$$

Mechanisms and Geometrical Non-linearity

If $\alpha_s>0$ the structure is statically indeterminate.

If $\alpha_s=0$ the structure is statically determinate.

If $\alpha_s<0$ the structure is a mechanism.

When choosing releases in a complex structure having a large statical indeterminacy, it is possible to introduce mechanisms. The value of α_s for the released structure may be zero, but this, in itself, is not a sufficient check. It is necessary to consider parts of the structure as complete structures in themselves, making the rest of the structure infinitely rigid.

Another possible source of trouble is the selection of releases which produce a non-linear statically determinate structure.

Consider, for example, the plane frame shown in Fig. 9 which has been made statically determinate by two different release systems, A and B.

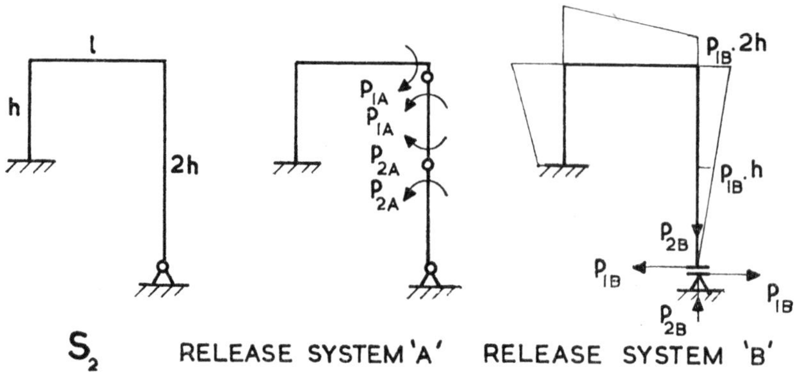

FIG. 9

Expressing the redundancies of the doubtful release system A in terms of the redundancies of the known system B, we obtain

$$\boldsymbol{p}_A = \boldsymbol{Q} \cdot \boldsymbol{p}_B$$

i.e.

$$\begin{bmatrix} p_{1A} \\ p_{2A} \end{bmatrix} = \begin{bmatrix} 2h & 0 \\ h & 0 \end{bmatrix} \begin{bmatrix} p_{1B} \\ p_{2B} \end{bmatrix}$$

$$\therefore \quad |Q| = 0$$

If the determinant of the matrix $\boldsymbol{Q}$ equals zero, then the release system A produces a mechanism and is, therefore, unacceptable.

Kinematical Indeterminacy

The kinematical indeterminacy of a structure is the number of possible relative displacements of the nodes in the directions of the stress resultants.

If there are N nodes and each node has six relative displacements relative to the adjacent nodes, then there are $6N$ degrees of freedom. However, the structure has six degrees of freedom as a rigid body and therefore

$$\alpha_k = 6(N-1)$$

A constraint exists where there is no possible relative displacement between two nodes or where the displacement does not create a stress resultant in the member connecting the nodes. For a structure containing c constraints and r releases, the kinematical indeterminacy becomes

$$\alpha_k = 6(N-1) - c + r \text{ for space structures}$$

$$\alpha_k = 3(N-1) - c + r \text{ for plane structures}$$

Consider the examples in Fig. 2

Fig. 2(a) Nodes A and E are fixed to the ground and therefore have no degrees of freedom. Nodes B, C and D each have three degrees of freedom in the plane of the frame. Hence, the total number of degrees of freedom of the nodes is nine. Alternatively, taking the ground as a single node

$$\alpha_k = 3(N-1) = 3(4-1) = 9$$

Considering A and E as separate nodes but remembering that there can be no relative movement between them i.e., there are three constraints, we have

$$\alpha_k = 3(5-1) - 3 = 9$$

Fig. 2(b) The upper node A has three degrees of freedom relative to the ground node B.

Alternatively $\alpha_k = 3(2-1) = 3$

Fig. 2(c) In the multi-storey frame, nodes B to K each have three degrees of freedom.

Therefore $\alpha_k = 10 \times 3 = 30$

Fig. 2(d) The nodes B to I are all part of the ground node and have no degrees of freedom. The node A has only two degrees of freedom i.e., two translations, since the rotation cannot produce any forces in a pin-jointed member.

Therefore $\alpha_k = 2$

Alternatively
$$\alpha_k = 3(N-1) - c + r$$
$$= 3(2-1) - 16 + 15 = 2$$

Fig. 2(e) In the pin-jointed plane truss, nodes A to M have each two degrees of freedom. Node N has zero degrees of freedom while nodes O and P have each got one degree of freedom making a total of $13 \times 2 + 2 \times 1 = 28$

Therefore $\alpha_k = 28$

Fig. 2(i) Nodes A, B and C have six degrees of freedom each. α_k is therefore $6 \times 3 = 18$.

Alternatively $\alpha_k = 6(4-1) = 18$ or $6(6-1) - 6 \times 2 = 18$

taking the ground nodes as 3 separate nodes.

If we ignore the axial deformations of the members, the degrees of freedom are reduced by 6 making $\alpha_k = 18 - 6 = 12$.

Consider now the introduction of ball-joints at each of the nodes D, E and F. The effect of this introduces three releases or three extra degrees of freedom at each of these nodes. The total degrees of freedom then become $18+3\times3=27$ or $12+3\times3=21$ if axial deformations are ignored.

Fig. 2(k) Nodes A, B and C each have three degrees of freedom since the rotations have no effect on the stress resultants in the members. The nodes D, E and F have no degrees of freedom of translation and therefore

$$\alpha_k=3\times3=9$$

THE FLEXIBILITY METHOD

Development of the Flexibility Influence Coefficients

A statically indeterminate structure may be made statically determinate by the provision of suitable releases, giving what is called the released structure.

Particular Solution

The stress resultants produced at any point in the released structure are in equilibrium with the applied loads. The column matrix $\boldsymbol{x}_0$ denotes the stress resultants due to a single applied loading system.

The number of elements in $\boldsymbol{x}_0$ is 1, 3 or 6 for structures classified under (i), (ii) and (iii) or (iv) respectively.

For space frames

$$\boldsymbol{x}_0=\begin{bmatrix} m_0 \\ s_0 \\ n_0 \\ g_0 \\ q_0 \\ t_0 \end{bmatrix}=\begin{bmatrix} x_{10} \\ x_{20} \\ x_{30} \\ x_{40} \\ x_{50} \\ x_{60} \end{bmatrix}$$

In general, $\boldsymbol{x}_0$ will be accompanied by discontinuities at each of the releases. These discontinuities are given by the column matrix $\boldsymbol{v}_0$.

$$\boldsymbol{v}_0=\begin{bmatrix} v_{10} \\ v_{20} \\ v_{30} \\ \cdot \\ \cdot \\ \cdot \\ v_{\alpha 0} \end{bmatrix}$$

Note the double suffix notation above. The first suffix denotes the type, while the second gives the cause. In this case zero for applied loading.

Complementary Solution

Consider bi-actions $\boldsymbol{p}$ acting at each of the releases with all the other loads removed, and producing stress resultants $\boldsymbol{x}$ at any point in the released structure. Each bi-action can produce up to six stress resultants at any point.

The stress resultants are denoted by the column matrix $\boldsymbol{x}$.

$$\underset{(6\times 1)}{\boldsymbol{x}} = \begin{bmatrix} x_1 \\ x_2 \\ x_3 \\ x_4 \\ x_5 \\ x_6 \end{bmatrix}$$

Expressing this in terms of the components produced by each of the bi-actions, we have

$$\underset{(6\times 1)}{\boldsymbol{x}} = \underset{(6\times\alpha)}{\boldsymbol{H}} \cdot \underset{(\alpha\times 1)}{\boldsymbol{p}} \tag{1}$$

$$= \begin{bmatrix} m_1 & m_2 & \ldots & m_\alpha \\ s_1 & s_2 & \ldots & s_\alpha \\ n_1 & n_2 & \ldots & n_\alpha \\ g_1 & g_2 & \ldots & g_\alpha \\ q_1 & q_2 & \ldots & q_\alpha \\ t_1 & t_2 & \ldots & t_\alpha \end{bmatrix} \begin{bmatrix} p_1 \\ p_2 \\ \cdot \\ \cdot \\ \cdot \\ p_\alpha \end{bmatrix}$$

where α is the number of bi-actions.

The components of $\boldsymbol{H}$ are the stress resultants due to *unit* bi-actions and $\boldsymbol{p}$ gives the unknown magnitudes of the bi-actions.

The stress resultants $\boldsymbol{x}$ acting on an element, *ds*, give rise to deformations expressed by the column matrix $\boldsymbol{du}$ of the same order as $\boldsymbol{x}$. Writing the total $\boldsymbol{dU}$ in terms of the components produced by each bi-action, we obtain

$$\boldsymbol{dU} = \begin{bmatrix} du_{11} & du_{12} & \ldots & du_{1\alpha} \\ du_{21} & du_{22} & \ldots & du_{2\alpha} \\ \vdots & & & \\ du_{61} & du_{62} & \ldots & du_{6\alpha} \end{bmatrix}$$

where each element du_{ij} represents the incremental deformation across an element *ds*.

i gives the type of displacement ($i=1, 2 \ldots 6$)
and j refers to the cause ($j=1, 2 \ldots \alpha$)
($j=0$ refers to the loading in the particular solution)

The deformations of the elemental lengths, *ds*, are related to the stress resultants by the expression

$$\underset{(6\times\alpha)}{\boldsymbol{dU}} = \underset{(6\times 6)}{\boldsymbol{F}} \cdot \underset{(6\times\alpha)}{\boldsymbol{H}} \cdot ds \tag{2}$$

where *ds* is a scalar element

i.e.

$$\begin{bmatrix} du_{11} & du_{12} & \ldots & du_{1\alpha} \\ du_{21} & du_{22} & \ldots & du_{2\alpha} \\ \cdot & \cdot & & \cdot \\ \cdot & \cdot & & \cdot \\ \cdot & \cdot & & \cdot \\ du_{61} & du_{62} & \ldots & du_{6\alpha} \end{bmatrix} = \begin{bmatrix} \frac{1}{EI_1} & 0 & 0 & 0 & 0 & 0 \\ 0 & \frac{1}{CA} & 0 & 0 & 0 & 0 \\ 0 & 0 & \frac{1}{EA} & 0 & 0 & 0 \\ 0 & 0 & 0 & \frac{1}{EI_2} & 0 & 0 \\ 0 & 0 & 0 & 0 & \frac{K}{CA} & 0 \\ 0 & 0 & 0 & 0 & 0 & \frac{1}{CJ} \end{bmatrix} \begin{bmatrix} m_1 & m_2 & \ldots & m_\alpha \\ s_1 & s_2 & \ldots & s_\alpha \\ n_1 & n_2 & \ldots & n_\alpha \\ g_1 & g_2 & \ldots & g_\alpha \\ q_1 & q_2 & \ldots & q_\alpha \\ t_1 & t_2 & \ldots & t_\alpha \end{bmatrix} ds$$

Now let us consider the relationship between the distortions, ***du***, of an element, *ds*, and the discontinuities at each of the releases produced by these distortions. Assuming that no other deformations but those of the element, *ds*, can occur, the members of the structure must move as a pure mechanism causing movements, ***dV***, at the releases. See Fig. 10

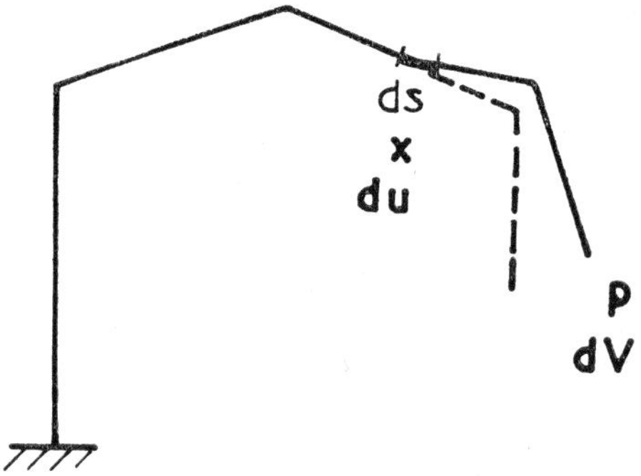

FIG. 10

Expressing this relationship in component form, we have

$$\underset{(\alpha\times\alpha)}{\begin{bmatrix} dv_{11} & dv_{12} & \ldots & dv_{1\alpha} \\ dv_{21} & dv_{22} & \ldots & dv_{2\alpha} \\ \vdots & \vdots & & \vdots \\ dv_{\alpha 1} & dv_{\alpha 2} & \ldots & dv_{\alpha\alpha} \end{bmatrix}} = \underset{(\alpha\times 6)}{\begin{bmatrix} k_{11} & k_{12} & \ldots & k_{16} \\ k_{21} & k_{22} & \ldots & k_{26} \\ \vdots & \vdots & & \vdots \\ k_{\alpha 1} & k_{\alpha 2} & \ldots & k_{\alpha 6} \end{bmatrix}} \underset{(6\times\alpha)}{\begin{bmatrix} du_{11} & du_{12} & \ldots & du_{1\alpha} \\ du_{21} & du_{22} & \ldots & du_{2\alpha} \\ \vdots & \vdots & & \vdots \\ du_{61} & du_{62} & \ldots & du_{6\alpha} \end{bmatrix}}$$

i.e.

$$\boldsymbol{dV}=\boldsymbol{K}\cdot\boldsymbol{dU}$$
$$=\boldsymbol{K}\cdot\boldsymbol{F}\cdot\boldsymbol{H}\cdot ds \tag{3}$$

where $\boldsymbol{K}$ is the Kinematical Matrix.

The work done by the stress resultants, $\boldsymbol{x}$, in causing the deformations, $\boldsymbol{du}$, of the element, ds, must equal the work done at the releases by the bi-actions, $\boldsymbol{p}$, acting through the displacements, $\boldsymbol{dv}$, assuming, of course, that no other deformations except those in element ds occur.

This gives

$$\boldsymbol{dv}^{\mathrm{T}}\cdot\boldsymbol{p}=\boldsymbol{du}^{\mathrm{T}}\cdot\boldsymbol{x} \tag{4}$$

where the column matrices $\boldsymbol{dv}$ and $\boldsymbol{du}$ have been transposed to conform with the multiplication.

$$\boldsymbol{dv}=\boldsymbol{dV}\cdot\boldsymbol{e} \quad \text{where} \quad \boldsymbol{e}=\text{Col. }(1,\ 1,\ 1\ \ldots\ 1)$$

In effect, postmultiplying $\boldsymbol{dV}$ by $\boldsymbol{e}$ sums up the component movements corresponding to each release.

$$\underset{(\alpha\times 1)}{\boldsymbol{dv}}=\begin{bmatrix} dv_1 \\ dv_2 \\ \vdots \\ dv_\alpha \end{bmatrix}=\underset{(\alpha\times\alpha)}{\boldsymbol{dV}}\cdot\underset{(\alpha\times 1)}{\boldsymbol{e}}=\begin{bmatrix} dv_{11} & dv_{12} & \ldots & dv_{1\alpha} \\ dv_{21} & dv_{22} & \ldots & dv_{2\alpha} \\ \vdots & \vdots & & \vdots \\ dv_{\alpha 1} & dv_{\alpha 2} & \ldots & dv_{\alpha\alpha} \end{bmatrix}\begin{bmatrix} 1 \\ 1 \\ \vdots \\ 1 \end{bmatrix}$$

$$=\begin{bmatrix} dv_{11}+dv_{12}+ & \ldots & dv_{1\alpha} \\ dv_{21}+dv_{22}+ & \ldots & dv_{2\alpha} \\ \vdots \quad \vdots & & \vdots \\ dv_{\alpha 1}+dv_{\alpha 2}+ & \ldots & dv_{\alpha\alpha} \end{bmatrix}$$

similarly

$$\underset{(6\times 1)}{\boldsymbol{du}}=\underset{(6\times\alpha)}{\boldsymbol{dU}}\cdot\underset{(\alpha\times 1)}{\boldsymbol{e}}$$

substituting in (4) we have

$$(\boldsymbol{dV}\cdot\boldsymbol{e})^{\mathrm{T}}\boldsymbol{p}=(\boldsymbol{dU}\cdot\boldsymbol{e})^{\mathrm{T}}\boldsymbol{x}$$
$$=(\boldsymbol{dU}\cdot\boldsymbol{e})^{\mathrm{T}}\boldsymbol{H}\cdot\boldsymbol{p}$$

but

$$\boldsymbol{dV}=\boldsymbol{K}\cdot\boldsymbol{dU}$$
$$\therefore\ (\boldsymbol{K}\cdot\boldsymbol{dU}\cdot\boldsymbol{e})^{\mathrm{T}}\boldsymbol{p}=(\boldsymbol{dU}\cdot\boldsymbol{e})^{\mathrm{T}}\boldsymbol{Hp}$$

taking the transpose of each side, and applying the reversal rule gives

$$\boldsymbol{p}^{\mathrm{T}}\cdot\boldsymbol{K}\cdot\boldsymbol{dU}\cdot\boldsymbol{e}=\boldsymbol{p}^{\mathrm{T}}\cdot\boldsymbol{H}^{\mathrm{T}}\cdot\boldsymbol{dU}\cdot\boldsymbol{e}$$

This can only be true if $\boldsymbol{K}=\boldsymbol{H}^{\mathrm{T}}$ numerically (5)

$$\therefore\ \boldsymbol{dV}=\boldsymbol{H}^{\mathrm{T}}\cdot\boldsymbol{F}\cdot\boldsymbol{H}\,ds \tag{6}$$

$\boldsymbol{H}^{\mathrm{T}}$, $\boldsymbol{F}$ and $\boldsymbol{H}$ are all functions of position on the structure.

The complete discontinuities caused by unit bi-actions at the releases can now be summed to give the matrix

$$\boldsymbol{V}=\int d\boldsymbol{V}=\int \boldsymbol{H}^{\mathrm{T}}\cdot\boldsymbol{F}\cdot\boldsymbol{H}\,ds \tag{7}$$

i.e.

$$\begin{bmatrix} v_{11} & v_{12} & \cdots & v_{1\alpha} \\ v_{21} & v_{22} & \cdots & v_{2\alpha} \\ \cdot & \cdot & & \cdot \\ \cdot & \cdot & & \cdot \\ \cdot & \cdot & & \cdot \\ v_{\alpha 1} & v_{\alpha 2} & \cdots & v_{\alpha\alpha} \end{bmatrix} = \int \begin{bmatrix} m_1 & s_1 & n_1 & g_1 & q_1 & t_1 \\ m_2 & s_2 & n_2 & g_2 & q_2 & t_2 \\ \cdot & \cdot & \cdot & \cdot & \cdot & \cdot \\ \cdot & \cdot & \cdot & \cdot & \cdot & \cdot \\ \cdot & \cdot & \cdot & \cdot & \cdot & \cdot \\ m_\alpha & s_\alpha & n_\alpha & g_\alpha & q_\alpha & t_\alpha \end{bmatrix} \begin{bmatrix} \frac{1}{EI_1} & 0 & 0 & 0 & 0 & 0 \\ 0 & \frac{1}{CA} & 0 & 0 & 0 & 0 \\ 0 & 0 & \frac{1}{EA} & 0 & 0 & 0 \\ 0 & 0 & 0 & \frac{1}{EI_2} & 0 & 0 \\ 0 & 0 & 0 & 0 & \frac{K}{CA} & 0 \\ 0 & 0 & 0 & 0 & 0 & \frac{1}{CJ} \end{bmatrix} \times$$

$$\begin{bmatrix} m_1 & m_2 & \cdots & m_\alpha \\ s_1 & s_2 & \cdots & s_\alpha \\ n_1 & n_2 & \cdots & n_\alpha \\ g_1 & g_2 & \cdots & g_\alpha \\ q_1 & q_2 & \cdots & q_\alpha \\ t_1 & t_2 & \cdots & t_\alpha \end{bmatrix} ds$$

The matrix $\boldsymbol{V}$ is called the FLEXIBILITY MATRIX of the complete structure and its elements may be expressed in general form as

$$v_{rs}=\int\frac{m_r m_s\,ds}{EI_1}+\int\frac{s_r s_s\,ds}{CA}+\int\frac{n_r n_s\,ds}{EA}+\int\frac{g_r g_s\,ds}{EI_2}+\int\frac{q_r q_s\,ds}{CA/K}+\int\frac{t_r t_s\,ds}{CJ} \tag{8}$$

These are the FLEXIBILITY INFLUENCE COEFFICIENTS which represent the discontinuities at the releases produced by unit bi-actions.

v_{rs} represents the movement in the direction r due to the unit force s.

Similarly, the particular solution yields a corresponding set of equations

$$d\boldsymbol{v}_0=\boldsymbol{K}\cdot d\boldsymbol{u}_0$$

and

$$d\boldsymbol{u}_0=\boldsymbol{F}\cdot\boldsymbol{x}_0\,ds$$

$$\therefore\quad d\boldsymbol{v}_0=\boldsymbol{K}\cdot\boldsymbol{F}\cdot\boldsymbol{x}_0\,ds$$

$$=\boldsymbol{H}^{\mathrm{T}}\cdot\boldsymbol{F}\cdot\boldsymbol{x}_0\,ds \tag{9}$$

The complete discontinuities, caused by one single loading system, can now be summed by integrating over the entire structure.

$$\boldsymbol{v}_0 = \int d\boldsymbol{v}_0 = \int \boldsymbol{H}^{\mathrm{T}} \cdot \boldsymbol{F} \cdot \boldsymbol{x}_0 \, ds \tag{10}$$

The total discontinuity at the releases in a released structure is, therefore, given by

$$\begin{aligned} \boldsymbol{v}_t &= \boldsymbol{v}_c + \boldsymbol{v}_0 \\ &= \boldsymbol{V} \cdot \boldsymbol{p} + \boldsymbol{v}_0 \end{aligned} \tag{11}$$

Generally, $\boldsymbol{v}_t = \boldsymbol{o}$ is the condition which restores continuity to the structure.

Therefore $$\boldsymbol{V} \cdot \boldsymbol{p} + \boldsymbol{v}_0 = \boldsymbol{o} \tag{12}$$

or $$\boldsymbol{p} = -\boldsymbol{V}^{-1} \cdot \boldsymbol{v}_0 \tag{13}$$

This is the equation of compatibility.

The total solution, which is the sum of the complementary solution and the particular solution, and which satisfies the conditions of continuity expressed by equation (12) is given by the expression

$$\boldsymbol{x}_t = \boldsymbol{x}_c + \boldsymbol{x}_0$$

or $$\boldsymbol{x}_t = \boldsymbol{H} \cdot \boldsymbol{p} + \boldsymbol{x}_0 \tag{14}$$

Therefore, $\boldsymbol{x}_c$ modifies $\boldsymbol{x}_0$ to restore continuity in the structure.

Summary of procedure for the flexibility method

Basic outline

1. Calculate α_s
2. Make S_α statically determinate by providing suitable releases.
3. Draw the P.S. diagrams ($\boldsymbol{x}_0$)
4. Draw the C.S. diagrams ($\boldsymbol{H}^{\mathrm{T}}$)
5. Evaluate the Flexibility Influence Coefficients i.e., the elements of $\boldsymbol{V}$ and $\boldsymbol{v}_0$
6. Compute $\boldsymbol{V}^{-1}$
7. Obtain $\boldsymbol{p}$ from $\boldsymbol{p} = -\boldsymbol{V}^{-1} \cdot \boldsymbol{v}_0$
8. Obtain $\boldsymbol{x}_t$ from $\boldsymbol{x}_t = \boldsymbol{H} \cdot \boldsymbol{p} + \boldsymbol{x}_0$ in tabular form
9. Apply checks.

The following detailed procedure is given to assist the reader in solving the problems in an orderly and correct manner.

Detailed procedure

1. In most structural problems it is possible to tell the degree of statical indeterminacy at a glance with a little experience. One merely counts the number of 'rings' or loops in the structure,

multiplies by 3 or 6 as appropriate and subtracts the number of releases present. If in any doubt, work out α_s from $\alpha_s = \frac{6}{3}(M-N+1) - r$

2. The number of releases required is given by α_s. Care must be taken in selecting releases in order to avoid mechanisms or non-linear behaviour conditions.
   (a) In general, cut releases are simpler and lead to simpler statics when working out the C.S. and P.S. diagrams.
   (b) If the structure is symmetrical, it is best to make the releases symmetrical. This often leads to zero elements of V and simplifies the matrix inversion. For example, consider a symmetrical portal frame shown in Fig. 11(a)

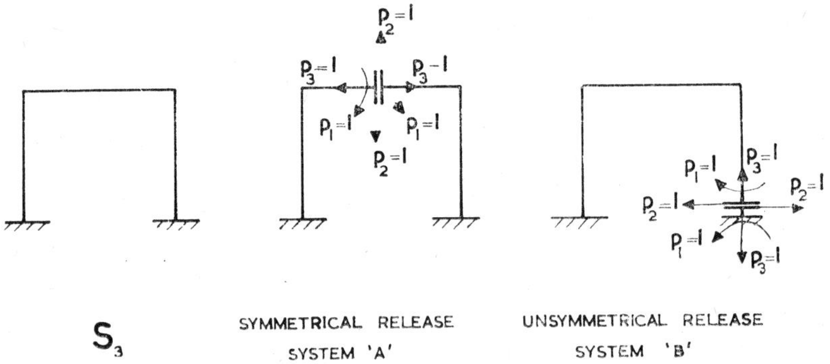

FIG. 11(a)

$$V_A = \begin{bmatrix} v_{11} & 0 & v_{13} \\ 0 & v_{22} & 0 \\ v_{31} & 0 & v_{23} \end{bmatrix} \qquad V_B = \begin{bmatrix} v_{11} & v_{12} & v_{13} \\ v_{21} & v_{22} & v_{23} \\ v_{31} & v_{32} & v_{33} \end{bmatrix}$$

Flexibility Matrix System A. Flexibility Matrix System B.

Considering bending energies only, the C.S. for System A is

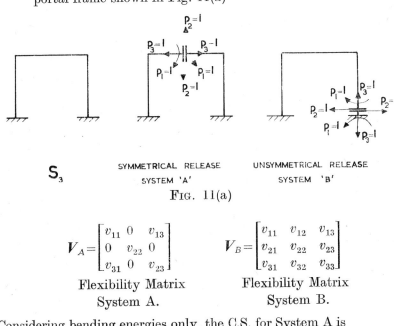

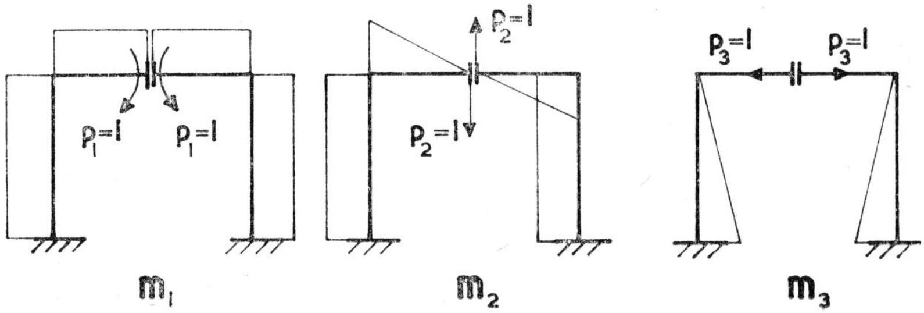

FIG. 11(b)

Owing to the symmetry and anti-symmetry, the flexibility coefficients

$$v_{12}=v_{21}=\int \frac{m_1 m_2\, ds}{EI}=0 \qquad \text{and} \qquad v_{23}=v_{32}=\int \frac{m_2 m_3\, ds}{EI}=0$$

(c) In a structure composed of a number of 'rings' or loops, choose a release system which localises the influence of the unit bi-actions. For example, consider a four-span continuous beam shown in Fig. 12

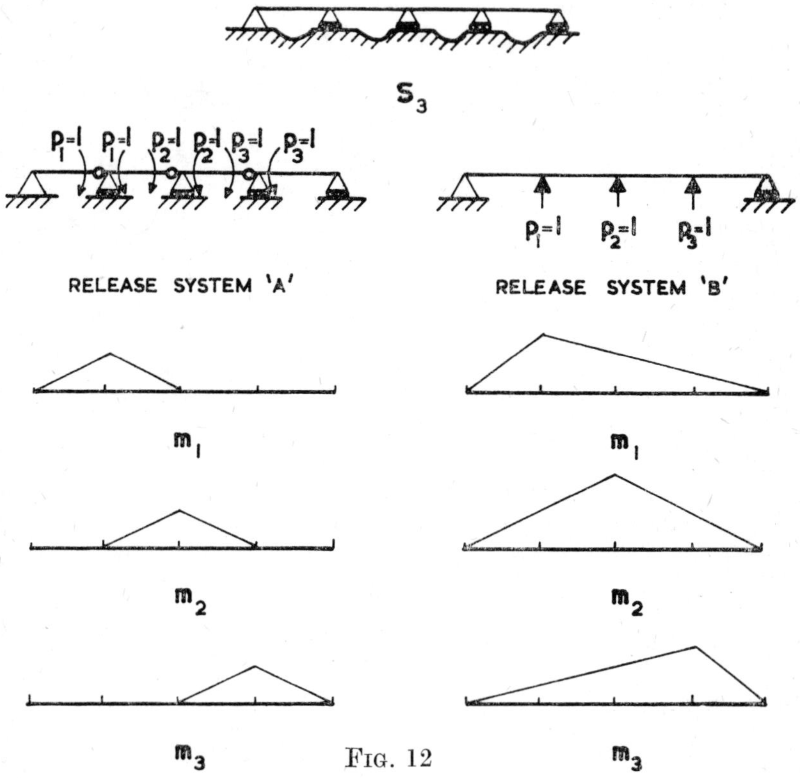

FIG. 12

For System A

$$V_A=\begin{bmatrix} v_{11} & v_{12} & 0 \\ v_{21} & v_{22} & v_{23} \\ 0 & v_{32} & v_{33} \end{bmatrix}$$

Matrix V_A contains zero elements and is better conditioned than Matrix V_B with the larger values forming the principal diagonal

For System B

$$V_B=\begin{bmatrix} v_{11} & v_{12} & v_{13} \\ v_{21} & v_{22} & v_{23} \\ v_{31} & v_{32} & v_{33} \end{bmatrix}$$

Matrix V_B contains no zero elements, and the small difference between large numbers makes the matrix inversion difficult.

(d) Making use of the 'elastic centre'

The 'elastic centre' is the centre of gravity of a structure whose weight is taken to be proportional to $\frac{1}{EI}$. This technique is applicable to the closed-ring type of structure and leads to a diagonal flexibility matrix which is easily inverted. For example see Fig. 13 below. Rigid arms leading from cut to elastic centre.

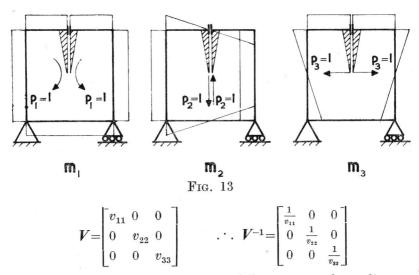

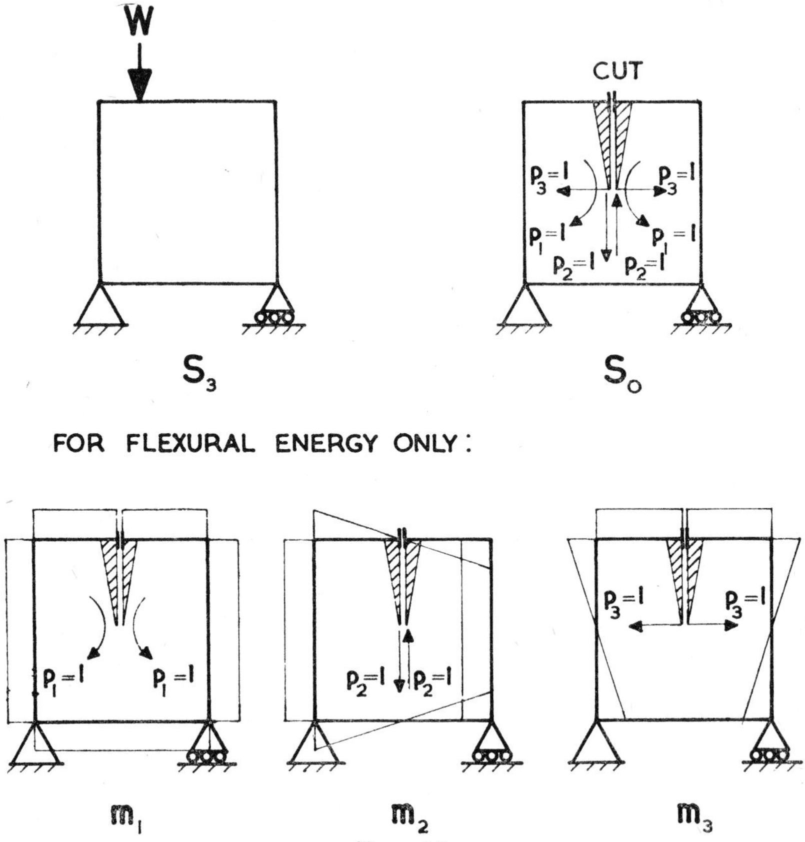

FIG. 13

$$V = \begin{bmatrix} v_{11} & 0 & 0 \\ 0 & v_{22} & 0 \\ 0 & 0 & v_{33} \end{bmatrix} \qquad \therefore \; V^{-1} = \begin{bmatrix} \frac{1}{v_{11}} & 0 & 0 \\ 0 & \frac{1}{v_{22}} & 0 \\ 0 & 0 & \frac{1}{v_{33}} \end{bmatrix}$$

3. The Particular Solution consists of the stress resultant diagrams produced by the action of applied loads on the released structure. These are obtained by simple statics i.e., by resolving forces and

taking moments. Although this is a simple part of the analysis, it is often the one which causes considerable difficulties because people have either forgotten the basic principles or never really understood them. Some simple exercises in statics are given in Fig. 15 for the benefit of the reader.

The following sign conventions will be adopted:

(a) *Bending Moments* (*m and g*). Draw the Bending Moment on the tension side of the member.

(b) *Shear Force* (*s and q*). To draw the shear force associated with the bending moment, put a direction arrow on each member and draw the shear force (Fig. 14) on 'top' of member when $\frac{dM}{dx}$ is positive in the direction of the arrow.

For example

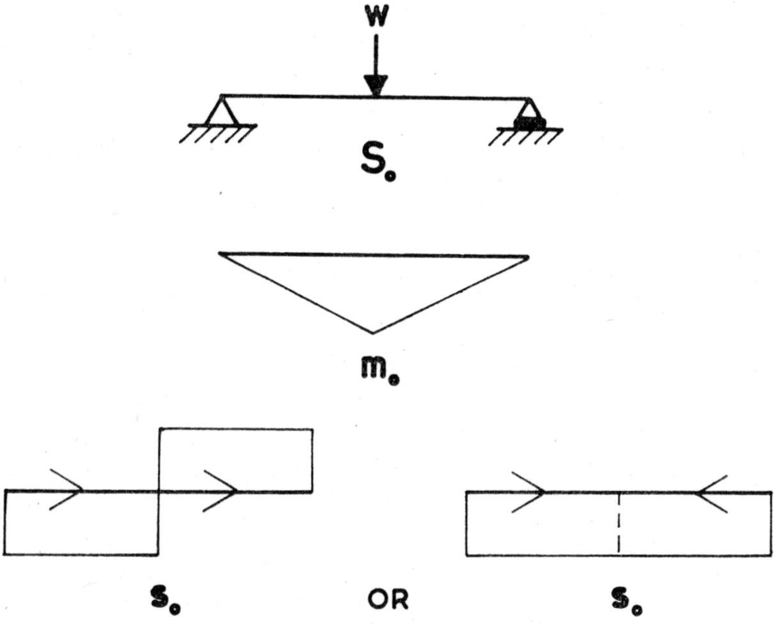

FIG. 14

It is worth remembering that for a constant Bending Moment, the shear force is zero since

$$\text{Shear Force} = \frac{dM}{dx} = 0$$

(c) *Direct Force* (*n*). Designate Tension $+ve$ and compression $-ve$.

EXERCISES SKETCH THE BENDING MOMENT, SHEAR FORCE AND AXIAL FORCE DIAGRAMS FOR THE FOLLOWING STATICALLY DETERMINATE SYSTEMS

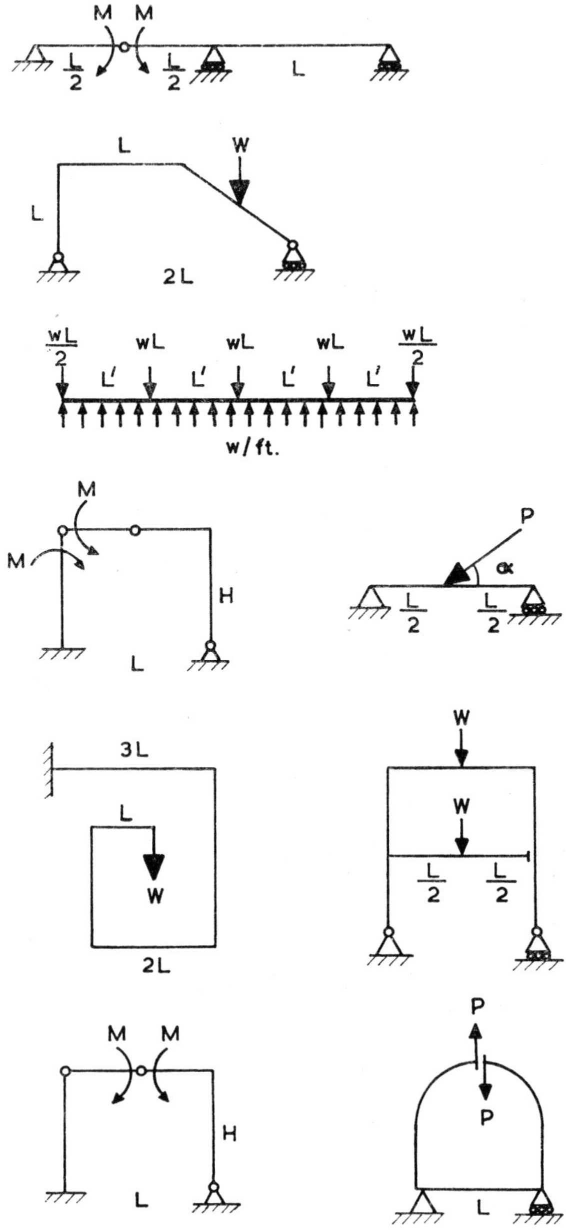

FIG. 15

(d) *Torsion* (t). Put a direction arrow on each member and draw the torsion as $+ve$ when the portion in front and viewed along the same direction as the arrow (Fig. 16) tends to twist in a clockwise direction.

For example

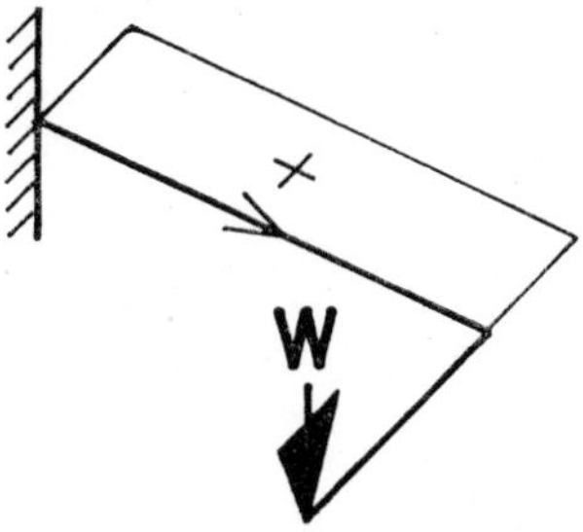

FIG. 16

4. The procedure for the C.S. diagrams is the same as for the P.S. except that the 'forces' become the unit bi-actions applied at the releases instead of the applied loads.

It is essential to draw all the stress resultant C.S. diagrams for which the total stress resultants are required, although only some of them are actually used in deriving the influence coefficients.

5. The general equation (8) for the Flexibility Coefficients is

$$v_{rs} = \int \frac{m_r m_s \, ds}{EI_1} + \int \frac{s_r s_s \, ds}{CA} + \int \frac{n_r n_s \, ds}{EA} + \int \frac{g_r g_s \, ds}{EI_2} + \int \frac{q_r q_s \, ds}{CA/K} + \int \frac{t_r t_s \, ds}{CJ}$$

where $r = 1$ to α and $s = 0$ to α.

0 represents the P.S.

The number of elements of $\boldsymbol{V}$ to be calculated is $\frac{\alpha}{2}(\alpha+1)$ since $v_{rs} = v_{sr}$ in a symmetric matrix.

Only the energy terms which contribute significantly to the deformation of the structure need to be considered in evaluating v_{rs}. For example, in a plane portal frame, the only significant energy is generally the bending energy and v_{rs} becomes

$$v_{rs} = \int \frac{m_r m_s \, ds}{EI}$$

In a plane grid with loads applied perpendicular to the plane of the grid, both g and t are significant and v_{rs} becomes

$$v_{rs} = \int \frac{g_r g_s \, ds}{EI} + \int \frac{t_r t_s \, ds}{CJ}$$

If in any doubt, the reader will soon realise what energies are significant when he evaluates the terms of equation (8).

The 'integration' is most conveniently carried out by means of Simpson's Rule or by using Table II which gives values for rapid computation.

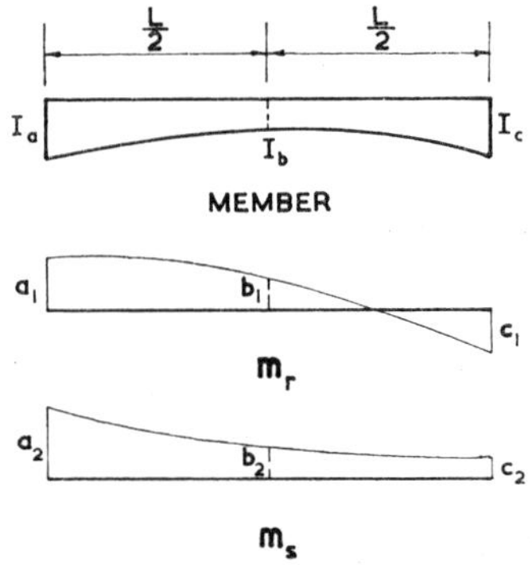

FIG. 17

By Simpson's Rule for flexural energy only, from Fig. 17

$$v_{rs}=\int\frac{m_r m_s\,ds}{EI}=\frac{l}{2\times 3E}\left[\frac{a_1a_2}{I_a}+\frac{4b_1b_2}{I_b}+\frac{c_1c_2}{I_c}\right]$$

For a prismatic member, i.e., $I=$Constant

$$v_{rs}=\frac{l}{2\times 3EI}\left[a_1a_2+4b_1b_2+c_1c_2\right]$$

where a_1, a_2, b_1, b_2 etc., all have their appropriate signs.

The flexibility influence coefficients are now assembled to form the flexibility matrix $\boldsymbol{V}$ for the structure and the column matrix $\boldsymbol{v}_0$.

6. The matrix $\boldsymbol{V}$ is inverted by one of the methods described in Part I.

7 & 8. The values of the unknowns p_1 to p_α are now evaluated from $\boldsymbol{p}=-\boldsymbol{V}^{-1}\cdot\boldsymbol{v}_0$ and the values of the total stress resultants obtained in a tablar form from the values of $\boldsymbol{p}$ and the C.S. and P.S. diagrams in the expression

$$\boldsymbol{x}_t=\boldsymbol{H}\cdot\boldsymbol{p}+\boldsymbol{x}_0$$

It should be noted that the flexibility matrix is independent of the loading, and depends only on the geometry of the structure, and the position of the releases provided. The structure can, therefore, be analysed for various applied loadings by evaluating the column matrix $\boldsymbol{v}_0$ for each case, and the corresponding values of $\boldsymbol{p}$ from $\boldsymbol{p}=-\boldsymbol{V}^{-1}\cdot\boldsymbol{v}_0$

TABLE II

$\int x_r x_s \, ds$						
x_r / x_s	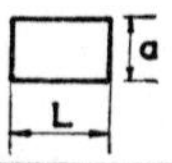	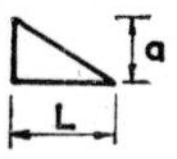				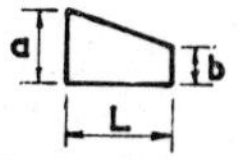
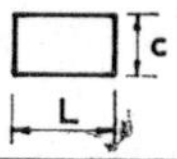	Lac					
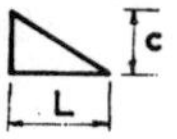	$\frac{Lac}{2}$	$\frac{Lac}{3}$				
	$\frac{Lac}{2}$	$\frac{Lac}{6}$	$\frac{Lac}{3}$			
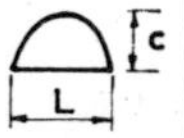	$\frac{2Lac}{3}$	$\frac{Lac}{3}$	$\frac{Lac}{3}$	$\frac{8Lac}{15}$		
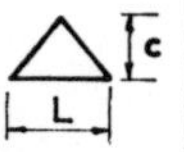	$\frac{Lac}{2}$	$\frac{Lac}{4}$	$\frac{Lac}{4}$	$\frac{5Lac}{12}$	$\frac{Lac}{3}$	
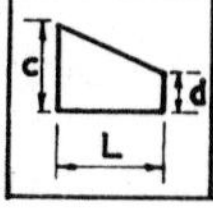	$\frac{La(c+d)}{2}$	$\frac{La(2c+d)}{6}$	$\frac{La(c+2d)}{6}$	$\frac{La(c+d)}{3}$	$\frac{La(c+d)}{4}$	$\frac{L}{6}\left[a(2c+d)+b(2d+c)\right]$

9. *Checks.* No analysis is complete without a systematic check. From equation (12) we have $\boldsymbol{V}\cdot\boldsymbol{p}+\boldsymbol{v}_0=\boldsymbol{o}$
i.e.

$$\int \boldsymbol{H}^{\mathrm{T}}\cdot\boldsymbol{F}\cdot\boldsymbol{H}\cdot\boldsymbol{p}\,ds+\int \boldsymbol{H}^{\mathrm{T}}\cdot\boldsymbol{F}\cdot\boldsymbol{x}_0\,ds=\boldsymbol{o}$$

or

$$\int \boldsymbol{H}^{\mathrm{T}}\cdot\boldsymbol{F}(\boldsymbol{H}\cdot\boldsymbol{p}+\boldsymbol{x}_0)\,ds=\boldsymbol{o} \qquad \text{But} \quad \boldsymbol{H}\cdot\boldsymbol{p}+\boldsymbol{x}_0=\boldsymbol{x}_t$$

$$\therefore \int \boldsymbol{H}^{\mathrm{T}}\cdot\boldsymbol{F}\cdot\boldsymbol{x}_t\,ds=\boldsymbol{o}=(\boldsymbol{\delta v}_0) \tag{15}$$

Thus, the result of 'integrating' the total stress resultant diagrams $\boldsymbol{x}_t$ with a set of C.S. diagrams $\boldsymbol{H}^{\mathrm{T}}$, should be zero. If the residuals $\boldsymbol{\delta v}_0$ obtained from equation (15) are very small compared with the column matrix $\boldsymbol{v}_0$ then the answer can be taken as correct.

It is essential to apply α checks corresponding to each of the α check releases. In applying these checks it is desirable to use a different release system to the one used for the C.S. and P.S. in the analysis.

In the case of a one times statically indeterminate structure S_1 the above check procedure is not satisfactory. All the possible C.S. diagrams are identical, and a complete check would not be provided. The method, in this case, consists of calculating the deflection at any chosen point by considering two different release systems. An illustration of this is given in example C.

As an additional check, the slope at any point on the final bending moment diagram m_t should give the value of the final shear force s_t at that point.

Calculation of Displacements

Displacements may be in the form of translations or rotations.

In equation (8) v_{rs} represents the displacements in direction r due to force s.

It follows, therefore, that the general equation for displacement in the direction δ due to the total stress resultants $\boldsymbol{x}_t$ becomes

$$\underset{(6\times 1)}{\boldsymbol{\delta}} = \int \underset{(6\times 6)}{\boldsymbol{H}_\delta} \cdot \underset{(6\times 6)}{\boldsymbol{F}} \cdot \underset{(6\times 1)}{\boldsymbol{x}_t}\, ds$$

where the displacement in any one direction $= \delta$

$$\delta = v_{\delta t} = \int \frac{m_\delta m_t\, ds}{EI} + \int \frac{s_\delta s_t\, ds}{CA} + \dots \int \frac{t_\delta t_t\, ds}{CJ} \tag{16}$$

Where m_δ, s_δ, n_δ etc. represent stress resultants due to unit load $p=1$ acting on the released structure in the direction of δ.

Procedure:

Only the energy terms which are significant in causing the displacements need be considered.

1. To evaluate the *deflection* at any point of a structure, apply a unit load $p=1$ to the *released* structure in the direction of the desired deflection, and draw the relevant stress resultant diagrams m_δ, s_δ, n_δ etc.

   The deflection is obtained from the product integrals of m_δ etc. and the total stress resultant diagrams m_t etc. given by equation (16).

   For a typical example, see the check part of the worked example C.

2. To evaluate the *rotation* at any point of a structure, apply a unit moment $p=1$ and draw the relevant stress resultant diagrams. The procedure is then the same as for deflections.

It should be noted that the above procedure for calculating displacements also applies to statically determinate structures.

Worked Examples:

Example A.

Obtain the stress resultant diagrams for the plane frame shown in Fig. 18.

Solution

Moment of inertia of Columns $=\frac{15\cdot18^3}{12}=7{,}290$ ins^4

Moment of inertia of beams $=\frac{15\cdot30^3}{12}=33{,}750$ ins^4

Ratio of Inertias $=\frac{33{,}750}{7{,}290}=4\cdot6296$

$$\therefore \quad EI_{\text{COL}}=1, \qquad EI_{\text{BEAM}}=4\cdot6296$$

The number of members, $M=7$

The number of nodes, $N=6$ (ground is a single mode)

The number of releases, $r=2$

$$\therefore \quad \alpha_s=3(7-6+1)-2=4$$

The structure is made statically determinate by the release system shown in Fig. 18.

The elements of the flexibility matrix V are derived from the complementary solution diagrams shown in Fig. 19.

Only the bending energy terms are included.

$$v_{11}=\tfrac{1}{1}\left[36\times36\times\tfrac{36}{3}\right]+\tfrac{1}{1}\left[21\times21\times\tfrac{21}{3}\right]+\tfrac{1}{4\cdot6296}[36\times36\times5]$$

$$+\tfrac{1}{4\cdot6296}\left[\tfrac{31\cdot3249}{2\times3}(36^2+4\times28\cdot5^2+21^2)\right]$$

$$=\underline{25{,}661\cdot42}$$

$$v_{12}=v_{21}=\tfrac{1}{1}\left[\tfrac{20}{2}(36+16)32\cdot5\right]+\tfrac{1}{4\cdot6296}\left[\tfrac{5}{2}(32\cdot5+27\cdot5)36\right]$$

$$+\tfrac{1}{4\cdot6296}\left[\tfrac{31\cdot3249}{3\times2}(72+21)27\cdot5\right]$$

$$=\underline{20{,}950\cdot51}$$

$$v_{13}=v_{31}=\tfrac{1}{1}\left[-\tfrac{20}{3\times2}(72+16)20\right]-\tfrac{1}{1}\left[\tfrac{5}{3\times2}(42+16)5\right]$$

$$-\tfrac{1}{4\cdot6296}[5\times20\times36]-\tfrac{1}{4\cdot6296}\left[\tfrac{31\cdot3249}{3\times2}([72+21]20+[36+42]5)\right]$$

$$=\underline{-9{,}423\cdot27}$$

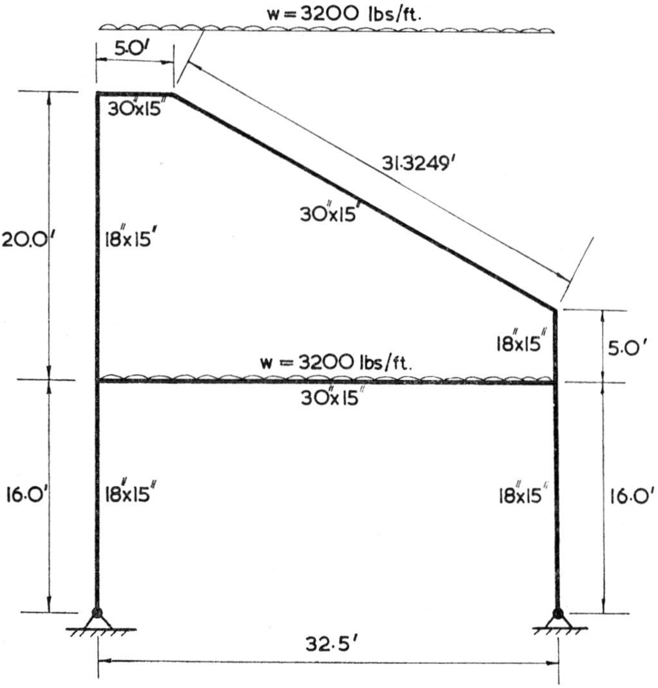

w = 3200 lbs/ft.
5.0′
30″x15″
31.3249′
30″x15″
20.0′
18″x15′
18″x15″
5.0′
w = 3200 lbs/ft.
30″x15″
16.0′
18″x15″
18″x15″
16.0′
32.5′

S_4

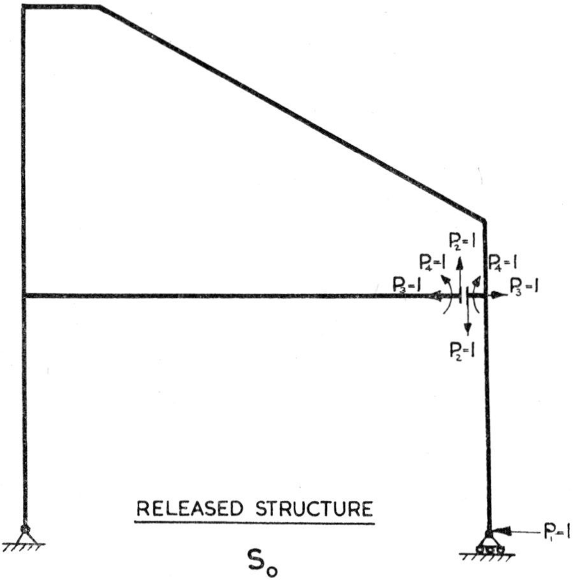

P2=1
P4=1
P4=1
P3=1
P3=1
P2=1
RELEASED STRUCTURE
P1=1

S_0

FIG. 18

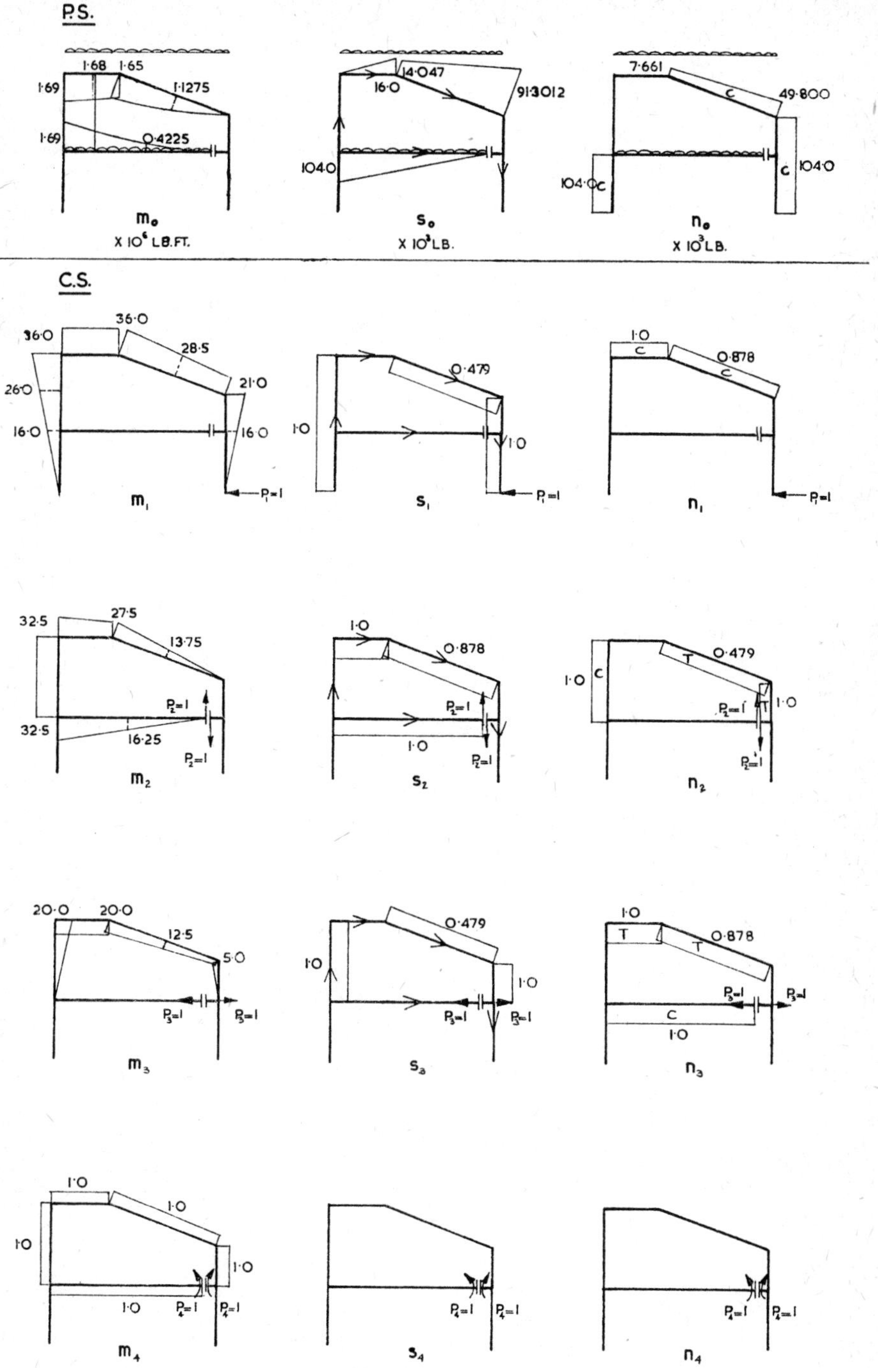
P.S.
1·68
1·65
1·69
1·1275
1·69
0·4225
m₀
X 10⁶ LB.FT.
14·047
16·0
91·3012
104·0
s₀
X 10³ LB.
7·661
C
49·800
104·0
C
104·0
n₀
X 10³ LB.
C.S.
36·0
36·0
28·5
26·0
21·0
16·0
16·0
P₁=1
m₁
0·479
1·0
1·0
P₁=1
s₁
1·0
0·878
C
C
P₁=1
n₁
32·5
27·5
13·75
P₂=1
32·5
16·25
P₂=1
m₂
1·0
0·878
P₂=1
1·0
P₂=1
s₂
0·479
T
1·0
C
P₂=1
T
1·0
P₂=1
n₂
20·0
20·0
12·5
5·0
P₃=1
P₃=1
m₃
0·479
1·0
1·0
P₃=1
P₃=1
s₃
1·0
T
0·878
T
P₃=1
P₃=1
C
1·0
n₃
1·0
1·0
1·0
1·0
1·0
P₄=1
P₄=1
m₄
P₄=1
P₄=1
s₄
P₄=1
P₄=1
n₄

FIG. 19

$$v_{14} = v_{41} = \frac{1}{1}\left[\frac{20}{2}(36+16)\right] + \frac{1}{1}\left[\frac{5}{2}(21+16)\right] + \frac{1}{4{\cdot}6296}[5\times 36]$$

$$+ \frac{1}{4{\cdot}6296}\left[\frac{31{\cdot}3249}{2}(36+21)\right]$$

$$= \underline{844{\cdot}22}$$

$$v_{22} = \frac{1}{1}[32{\cdot}5\times 32{\cdot}5\times 20] + \frac{1}{4{\cdot}6296}\left[32{\cdot}5\times 32{\cdot}5\times \frac{32{\cdot}5}{3}\right]$$

$$+ \frac{1}{4{\cdot}6296}\left[\frac{5}{3\times 2}([65+27{\cdot}5]32{\cdot}5+[55+32{\cdot}5]27{\cdot}5)\right]$$

$$+ \frac{1}{4{\cdot}6296}\left[27{\cdot}5\times 27{\cdot}5\times \frac{31{\cdot}3249}{3}\right]$$

$$= \underline{26{,}276{\cdot}55}$$

$$v_{23} = v_{32} = -\frac{1}{1}\left[20\times 32{\cdot}5\times \frac{20}{2}\right] - \frac{1}{4{\cdot}6296}\left[\frac{5}{2}(32{\cdot}5+27{\cdot}5)20\right]$$

$$- \frac{1}{4{\cdot}6296}\left[\frac{31{\cdot}3249}{3\times 2}(40+5)27{\cdot}5\right]$$

$$= \underline{-8{,}543{\cdot}54}$$

$$v_{24} = v_{42} = \frac{1}{1}[20\times 32{\cdot}5] + \frac{1}{4{\cdot}6296}\left[\frac{5}{2}(32{\cdot}5+27{\cdot}5)\right]$$

$$+ \frac{1}{4{\cdot}6296}\left[\frac{31{\cdot}3249}{2}\times 27{\cdot}5\right] + \frac{1}{4{\cdot}6296}\left[\frac{32{\cdot}5}{2}\times 32{\cdot}5\right]$$

$$= \underline{889{\cdot}51}$$

$$v_{33} = \frac{1}{1}\left[\frac{20}{3}\times 20\times 20\right] + \frac{1}{1}\left[\frac{5}{3}\times 5\times 5\right] + \frac{1}{4{\cdot}6296}[5\times 20\times 20]$$

$$+ \frac{1}{4{\cdot}6296}\left[\frac{31{\cdot}3249}{3\times 2}([40+5]20+[20+10]5)\right]$$

$$= \underline{4{,}324{\cdot}43}$$

$$v_{34} = v_{43} = -\frac{1}{1}\left[\frac{20}{2}\times 20\right] - \frac{1}{1}\left[\frac{5}{2}\times 5\right] - \frac{1}{4{\cdot}6296}[5\times 20]$$

$$- \frac{1}{4{\cdot}6296}\left[\frac{31{\cdot}3249}{2}(20+5)\right]$$

$$= \underline{-318{\cdot}68}$$

$$v_{44} = \frac{1}{1}[20] + \frac{1}{1}[5] + \frac{1}{4{\cdot}6296}[32{\cdot}5+5+31{\cdot}3249]$$

$$= \underline{39{\cdot}87}$$

The inverse of the flexibility matrix V must now be calculated. The procedure is as outlined in Part I.

$$V = \begin{bmatrix} 25{,}661{\cdot}42 & 20{,}950{\cdot}51 & -9{,}423{\cdot}27 & 844{\cdot}22 \\ 20{,}950{\cdot}51 & 26{,}276{\cdot}55 & -8{,}543{\cdot}54 & 889{\cdot}51 \\ -9{,}423{\cdot}27 & -8{,}543{\cdot}54 & 4{,}324{\cdot}43 & -318{\cdot}68 \\ 844{\cdot}22 & 889{\cdot}51 & -318{\cdot}68 & 39{\cdot}87 \end{bmatrix} \begin{matrix} S_1 \\ 38{,}032{\cdot}88 \\ 39{,}573{\cdot}03 \\ -13{,}961{\cdot}06 \\ 1{,}454{\cdot}92 \end{matrix}$$

$(C \times C)$

$$L^{\mathrm{T}} = \begin{bmatrix} 160{\cdot}191823 & 130{\cdot}783892 & -58{\cdot}824913 & 5{\cdot}270057 \\ 0 & 95{\cdot}771205 & -8{\cdot}877292 & 2{\cdot}091145 \\ 0 & 0 & 28{\cdot}022371 & 0{\cdot}353088 \\ 0 & 0 & 0 & 2{\cdot}756618 \end{bmatrix} \begin{matrix} S_2 \\ 237{\cdot}420859\ \checkmark \\ 88{\cdot}985058\ \checkmark \\ 28{\cdot}375459\ \checkmark \\ 2{\cdot}756618\ \checkmark \end{matrix}$$

$(R \times R)$

diag $L^{-1} = \{0{\cdot}00624252 \quad 0{\cdot}01044155 \quad 0{\cdot}03568577 \quad 0{\cdot}36276336\}$

$$V^{-1} = \begin{bmatrix} 0{\cdot}000266116 & 0{\cdot}000002153 & 0{\cdot}000402350 & -0{\cdot}002466860 \\ 0{\cdot}000002153 & 0{\cdot}000189590 & 0{\cdot}00015618 & -0{\cdot}00302709 \\ 0{\cdot}000402350 & 0{\cdot}00015618 & 0{\cdot}00129436 & -0{\cdot}00165815 \\ -0{\cdot}002466860 & -0{\cdot}00302709 & -0{\cdot}00165815 & 0{\cdot}13159725 \end{bmatrix}$$

The values of the final checks are given below.

$$\text{Col. 4 of } V^{-1} \times S_1 = 1{\cdot}00008888$$
$$\text{Col. 3 of } V^{-1} \times S_1 = 0{\cdot}99993187$$
$$\text{Col. 2 of } V^{-1} \times S_1 = 0{\cdot}99992341$$
$$\text{Col. 1 of } V^{-1} \times S_1 = 1{\cdot}00004867$$

The elements of the column matrix v_0 are derived from the particular and complementary solution diagrams shown in Fig. 19.

$$v_{10} = -\tfrac{1}{1}\left[\tfrac{20}{2}(16+36)1{\cdot}69\right]10^6 - \tfrac{1}{4{\cdot}6296}\left[\tfrac{5}{2}(1{\cdot}69+1{\cdot}65)36\right]10^6$$

$$\quad - \tfrac{1}{4{\cdot}6296}\left[\tfrac{31{\cdot}3249}{3\times2}(36\times1{\cdot}65+4\times28{\cdot}5\times1{\cdot}1275)\right]10^6$$

$$\quad = -1{,}155{,}664{,}995$$

$$v_{20} = -\tfrac{1}{1}[20\times32{\cdot}5\times1{\cdot}69]10^6$$

$$\quad - \tfrac{10^6}{4{\cdot}6296}\left[\tfrac{5}{3\times2}([3{\cdot}38+1{\cdot}65]32{\cdot}5+[1{\cdot}69+3{\cdot}3]27{\cdot}5)\right]$$

$$\quad - \tfrac{10^6}{4{\cdot}6296}\left[\tfrac{31{\cdot}3249}{3\times2}(1{\cdot}65\times27{\cdot}5+4\times13{\cdot}75\times1{\cdot}1275)\right]$$

$$\quad - \tfrac{10^6}{4{\cdot}6296}\left[\tfrac{32{\cdot}5}{3\times2}(1{\cdot}69\times32{\cdot}5+4\times0{\cdot}4225\times16{\cdot}25)\right]$$

$$\quad = -1{,}370{,}121{,}513$$

$$v_{30}=\frac{10^6}{1}\left[\frac{20}{2}\times 1{\cdot}69\times 20\right]+\frac{10^6}{4{\cdot}6296}\left[\frac{5}{2}(1{\cdot}69+1{\cdot}65)20\right]$$

$$+\frac{10^6}{4{\cdot}6296}\left[\frac{31{\cdot}3249}{3\times 2}(1{\cdot}65\times 20+4\times 1{\cdot}1275\times 12{\cdot}5)\right]$$

$$=\underline{474{,}860{,}742}$$

$$v_{40}=-\frac{10^6}{1}[20\times 1{\cdot}69]-\frac{1}{4{\cdot}6296}\left[\frac{32{\cdot}5}{3\times 2}(1{\cdot}69+4\times 0{\cdot}4225)\right]$$

$$-\frac{1}{4{\cdot}6296}\left[\frac{5}{2}(1{\cdot}69+1{\cdot}65)\right]-\frac{1}{4{\cdot}6296}\left[\frac{31{\cdot}3249}{3\times 2}(1{\cdot}65+4\times 1{\cdot}1275)\right]$$

$$=\underline{-46{,}504{,}891}$$

i.e.

$$\boldsymbol{v}_0=\begin{bmatrix} -1{,}155{,}664{,}995 \\ -1{,}370{,}121{,}513 \\ 474{,}860{,}742 \\ -\quad 46{,}504{,}891 \end{bmatrix}$$

$$\boldsymbol{V}\cdot\boldsymbol{p}+\boldsymbol{v}_0=\boldsymbol{o} \qquad \therefore \ \boldsymbol{p}=-\boldsymbol{V}^{-1}\cdot\boldsymbol{v}_0$$

Hence

$$\boldsymbol{p}=\begin{bmatrix} 4{,}709{\cdot}5425 \\ 47{,}311{\cdot}2432 \\ -12{,}785{\cdot}4464 \\ -91{,}038{\cdot}7739 \end{bmatrix}=\begin{bmatrix} p_1 \\ p_2 \\ p_3 \\ p_4 \end{bmatrix}$$

The total stress resultants $\boldsymbol{x}_t$ at every point on the structure are obtained from the particular and complementary solution diagrams and the expression

$$\boldsymbol{x}_t=\boldsymbol{H}\cdot\boldsymbol{p}+\boldsymbol{x}_0$$

i.e.

$$m_t=m_1p_1+m_2p_2+m_3p_3+m_4p_4+m_0$$

$$s_t=s_1p_1+s_2p_2+s_3p_3+s_4p_4+s_0$$

$$n_t=n_1p_1+n_2p_2+n_3p_3+n_4p_4+n_0$$

For convenience these values are calculated in tabular form. The total stress resultant diagrams are given in Fig. 20.

Checks

The values of the total stress resultants obtained are checked by using the release system shown in Fig. 21.

TOTAL MOMENTS

m_t lb−ft	= m_1p_1 $p_1=4{,}709{\cdot}5425$	+ m_2p_2 $p_2=47{,}311{\cdot}2432$	+ m_3p_3 $p_3=-12{,}785{\cdot}4464$	+ m_4p_4 $p_4=-91{,}038{\cdot}7739$	+ m_0	= m_t
m_t^{BA}	$-16{\cdot}0\ p_1$	—	—	—	–	$-\ 75{,}352{\cdot}6795$
m_t^{BC}	$-16{\cdot}0\ p_1$	$-32{\cdot}5\ p_2$	—	$-1{\cdot}0\ p_4$	+1,690,000	$+168{,}070{\cdot}6903$
m_t^{I}	$-26{\cdot}0\ p_1$	$-32{\cdot}5\ p_2$	$+10{\cdot}0\ p_3$	$-1{\cdot}0\ p_4$	+1,690,000	$-\ 6{,}879{\cdot}1982$
m_t^{C}	$-36{\cdot}0\ p_1$	$-32{\cdot}5\ p_2$	$+20{\cdot}0\ p_3$	$-1{\cdot}0\ p_4$	+1,690,000	$-181{,}829{\cdot}0868$
m_t^{J}	$-36{\cdot}0\ p_1$	$-30{\cdot}0\ p_2$	$+20{\cdot}0\ p_3$	$-1{\cdot}0\ p_4$	+1,680,000	$-\ 73{,}550{\cdot}9788$
m_t^{D}	$-36{\cdot}0\ p_1$	$-27{\cdot}5\ p_2$	$+20{\cdot}0\ p_3$	$-1{\cdot}0\ p_4$	+1,650,000	$+\ 14{,}727{\cdot}1292$
m_t^{K}	$-28{\cdot}5\ p_1$	$-13{\cdot}75\ p_2$	$+12{\cdot}5\ p_3$	$-1{\cdot}0\ p_4$	+1,127,500	$+273{,}969{\cdot}1396$
m_t^{E}	$-21{\cdot}0\ p_1$	—	$+\ 5{\cdot}0\ p_3$	$-1{\cdot}0\ p_4$	—	$-\ 71{,}788{\cdot}8500$
m_t^{L}	$-18{\cdot}5\ p_1$	—	$+\ 2{\cdot}5\ p_3$	$-1{\cdot}0\ p_4$	—	$-\ 28{,}051{\cdot}3778$
m_t^{FE}	$-16{\cdot}0\ p_1$	—	—	$-1{\cdot}0\ p_4$	—	$+\ 15{,}686{\cdot}0943$
m_t^{FG}	$-16{\cdot}0\ p_1$	—	—	—	—	$-\ 75{,}352{\cdot}6795$
m_t^{BF}	—	$+32{\cdot}5\ p_2$	—	$+1{\cdot}0\ p_4$	−1,690,000	$-243{,}423{\cdot}3699$
m_t^{H}	—	$+16{\cdot}25\ p_2$	—	$+1{\cdot}0\ p_4$	− 422,500	$+255{,}268{\cdot}9281$
m_t^{FB}	—	—	—	$+1{\cdot}0\ p_4$	—	$-\ 91{,}038{\cdot}7739$

TOTAL SHEAR FORCE

s_t lbs	= s_1p_1 $p_1=4{,}709{\cdot}5425$	+ s_2p_2 $p_2=47{,}311{\cdot}2432$	+ s_3p_3 $p_3=-12{,}785{\cdot}4464$	+ s_4p_4 $p_4=-91{,}038{\cdot}7739$	+ s_0	= s_t
s_t^{BA}	$-1{\cdot}0\ p_1$	—	—	—	—	$-\ 4{,}710$
s_t^{BC}	$-1{\cdot}0\ p_1$	—	$+1{\cdot}0\ p_3$	—	—	$-17{,}495$
s^I_t	$-1{\cdot}0\ p_1$	—	$+1{\cdot}0\ p_3$	—	—	$-17{,}495$
s_t^{CB}	$-1{\cdot}0\ p_1$	—	$+1{\cdot}0\ p_3$	—	—	$-17{,}495$
s_t^{CD}	—	$+1{\cdot}0\ p_2$	—	—	—	$+47{,}311$
s_t^J	—	$+1{\cdot}0\ p_2$	—	—	$-\ 8{,}000$	$+39{,}311$
s^{DC}	—	$+1{\cdot}0\ p_2$	—	—	$-\ 16{,}000$	$+31{,}311$
s_t^{DE}	$+0{\cdot}479\ p_1$	$+0{\cdot}878\ p_2$	$-0{\cdot}479\ p_3$	—	$-\ 14{,}047$	$+35{,}872$
s_t^K	$+0{\cdot}479\ p_1$	$+0{\cdot}878\ p_2$	$-0{\cdot}479\ p_3$	—	$-\ 52{,}674$	$-\ 2{,}755$
s_t^{ED}	$+0{\cdot}479\ p_1$	$+0{\cdot}878\ p_2$	$-0{\cdot}479\ p_3$	—	$-\ 91{,}301$	$-41{,}382$
s_t^{EF}	$+1{\cdot}0\ p_1$	—	$-1{\cdot}0\ p_3$	—	—	$+17{,}495$
s_t^L	$+1{\cdot}0\ p_1$	—	$-1{\cdot}0\ p_3$	—	—	$+17{,}495$
s_t^{FE}	$+1{\cdot}0\ p_1$	—	$-1{\cdot}0\ p_3$	—	—	$+17{,}495$
s_t^{FG}	$+1{\cdot}0\ p_1$	—	—	—	—	$+\ 4{,}710$
s_t^{BF}	—	$-1{\cdot}0\ p_2$	—	—	$+104{,}000$	$+56{,}689$
s_t^H	—	$-1{\cdot}0\ p_2$	—	—	$+\ 52{,}000$	$+\ 4{,}689$
s_t^{FB}	—	$-1{\cdot}0\ p_2$	—	—	—	$-47{,}311$

TOTAL AXIAL FORCE

n_t lbs	= $n_1 p_1$ + $p_1 = 4{,}709{\cdot}5425$	$n_2 p_2$ + $p_2 = 47{,}311{\cdot}2432$	$n_3 p_3$ + $p_3 = -12{,}785{\cdot}4464$	$n_4 p_4$ + $p_4 = -91{,}038{\cdot}7739$	n_0 =	n_t
n_t^{AB}	—	—	—	—	– 104,000	– 104,000
n_t^{BC}	—	$-1{\cdot}0\ p_2$	—	—	—	– 47,311
n_t^{CD}	$-1{\cdot}0\ p_1$	—	$+1{\cdot}0\ p_3$	—	—	– 17,495
n_t^{DK}	$-0{\cdot}878\ p_1$	$+0{\cdot}479\ p_2$	$+0{\cdot}878\ p_3$	—	– 7,661	– 360
n_t^{K}	$-0{\cdot}878\ p_1$	$+0{\cdot}479\ p_2$	$+0{\cdot}878\ p_3$	—	– 28,731	– 21,430
n_t^{EK}	$-0{\cdot}878\ p_1$	$+0{\cdot}479\ p_2$	$+0{\cdot}878\ p_3$	—	– 49,800	– 42,499
n_t^{EF}	—	$+1{\cdot}0\ p_2$	—	—	– 104,000	– 56,689
n_t^{FG}	—	—	—	—	– 104,000	– 104,000
n_t^{BF}	—	—	$-1{\cdot}0\ p_3$	—	—	+ 12,785

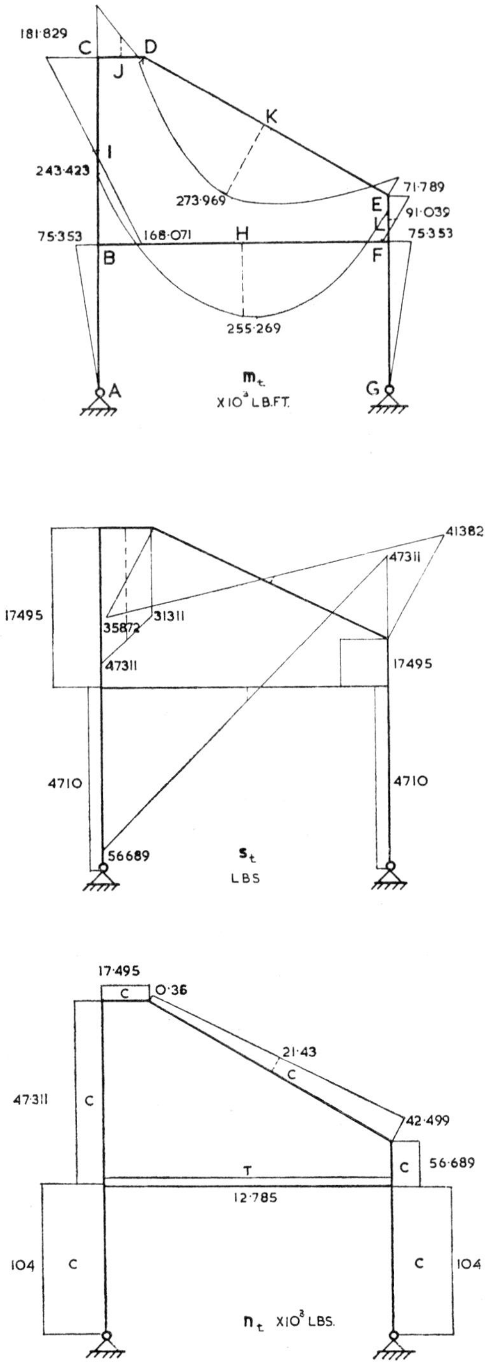
181·829
C
D
J
K
I
243·423
273·969
71·789
E
91·039
L
75·353
168·071
H
75·353
B
F
255·269
m_t
A
G
X10³ LB.FT.
41382
47311
17495
31311
35872
47311
17495
4710
4710
56689
s_t
LBS
17·495
C
0·36
21·43
C
47·311
C
42·499
T
C
56·689
12·785
104
C
C
104
n_t X10³ LBS.

FIG. 20

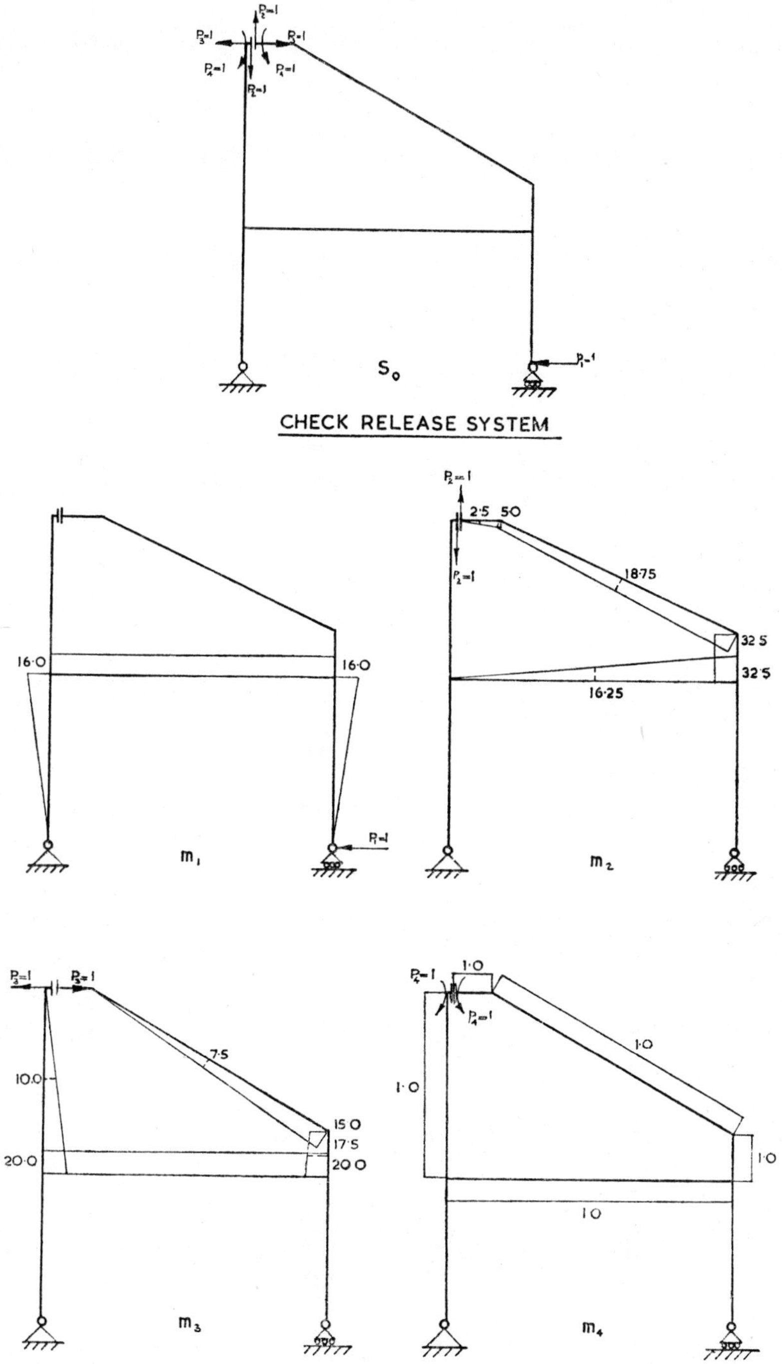

P2=1
P3=1
P4=1
P1=1
S0
P1=1
CHECK RELEASE SYSTEM
16·0
16·0
m1
P1=1
P2=1
2·5
5·0
18·75
32·5
32·5
16·25
m2
P3=1
10·0
7·5
15·0
17·5
20·0
20·0
m3
P4=1
1·0
1·0
1·0
1·0
1·0
m4

FIG. 21.

1st Check

$$\int \frac{m_1 m_t\, ds}{EI} = \delta v_1$$

$$= \frac{16}{3}[16\times 75{,}352\cdot 6795] + \frac{16}{3}[16\times 75{,}352\cdot 6795]$$

$$+ \frac{32\cdot 5\times 16}{6\times 4\cdot 6296}[243{,}423\cdot 3699 - 4\times 255{,}268\cdot 9281 + 91{,}038\cdot 7739]$$

$$= +\underline{6{,}702\cdot 37}$$

2nd Check

$$\int \frac{m_2 m_t\, ds}{EI} = \delta v_2$$

$$= \frac{32\cdot 5}{6\times 4\cdot 6296}[-4\times 16\cdot 25\times 255{,}268\cdot 9281 + 32\cdot 5\times 91{,}038\cdot 7739]$$

$$+ \frac{5\times 32\cdot 5}{6}[15{,}686\cdot 0943 - 4\times 28{,}051\cdot 3778 - 71{,}788\cdot 8500]$$

$$+ \frac{31\cdot 3249}{6\times 4\cdot 6296}[-32\cdot 5\times 71{,}788\cdot 85 + 4\times 18\cdot 75\times 273{,}969\cdot 1396 + 5\cdot 0\times 14{,}727\cdot 1292]$$

$$+ \frac{5\cdot 0}{6\times 4\cdot 6296}[5\cdot 0\times 14{,}727\cdot 1292 - 4\times 2\cdot 5\times 73{,}550\cdot 9788]$$

$$= -\underline{5{,}391\cdot 03}$$

3rd Check

$$\int \frac{m_3 m_t\, ds}{EI} = \delta v_3$$

$$= \frac{20}{6}[-4\times 10\times 6{,}879\cdot 1982 + 20\times 168{,}070\cdot 6903]$$

$$+ \frac{32\cdot 5\times 20}{6\times 4\cdot 6296}[243{,}423\cdot 3699 - 4\times 255{,}268\cdot 9281 + 91{,}038\cdot 7739]$$

$$+ \frac{5}{6}[20\times 15{,}686\cdot 0943 - 4\times 17\cdot 5\times 28{,}051\cdot 3778 - 15\times 71{,}788\cdot 8500]$$

$$+ \frac{31\cdot 3249}{6\times 4\cdot 6296}[-15\cdot 0\times 71{,}788\cdot 85 + 4\times 7\cdot 5\times 273{,}969\cdot 1396]$$

$$= +\underline{2{,}701\cdot 34}$$

4th Check

$$\int \frac{m_4 m_t\, ds}{EI} = \delta v_4$$

$$= \frac{20}{6}[181{,}829\cdot 0868 + 4\times 6879\cdot 1982 - 168{,}070\cdot 6903]$$

$$+ \frac{32\cdot 5}{6\times 4\cdot 6296}[-243{,}423\cdot 3699 + 4\times 255{,}268\cdot 9281 - 91{,}038\cdot 7739]$$

$$+ \frac{5}{6}[71{,}788\cdot 85 + 4\times 28{,}051\cdot 3778 - 15{,}686\cdot 0943]$$

$$+ \frac{31\cdot 3249}{6\times 4\cdot 6296}[71{,}788\cdot 85 - 4\times 273{,}969\cdot 1396 - 14{,}727\cdot 1292]$$

$$+ \frac{5\cdot 0}{6\times 4\cdot 6296}[-14{,}727\cdot 1292 + 4\times 73{,}550\cdot 9788 + 181{,}829\cdot 0868]$$

$$= -\underline{7{,}259\cdot 53}$$

The residuals are therefore

$$\boldsymbol{\delta v}_0 = \begin{bmatrix} 6{,}702{\cdot}37 \\ -5{,}391{\cdot}03 \\ 2{,}701{\cdot}34 \\ -7{,}259{\cdot}53 \end{bmatrix}$$

Compared with the values of $\boldsymbol{v}_0$ obtained earlier these are very small and therefore the answer may be assumed to be correct.

Example B.

Determine and draw the two bending moment and torque diagrams for the space frame shown in Fig. 22.

For all members take $EI_1 = EI_2 = 1$ and $CJ = \frac{1}{2}$.

Solution

The structure is three times statically indeterminate and is released by a cut at the ball joint at F. The released structure $\boldsymbol{S}_0$ is a three-dimensional cantilever. Only the strain energies due to the two perpendicular moments and the torque (i.e. m, g and t) are significant in causing the deformation of the structure.

To establish the moments in the plane of the frame (m), and moments normal to the plane of the frame (g), note that the members of the frame lie in two planes i.e. ABD and DEF

From the P.S. and C.S. diagrams we have therefore

$$v_{rs} = \int \frac{m_r m_s\, ds}{EI_1} + \int \frac{g_r g_s\, ds}{EI_2} + \int \frac{t_r t_s\, ds}{CJ}$$

which yields

$$v_{11} = \left[\frac{10\times 10\times 10}{1} + \frac{2(10\times 10\times 10)}{1\times 3}\right] + \left[\frac{10\times 10\times 10}{1}\right] + \left[\frac{10\times 10\times 10}{\frac{1}{2}}\right]$$

$$= 1000\left(\tfrac{14}{3}\right)$$

$$v_{12} = v_{21} = \left[\frac{-10\times 10\times 10}{1\times 2}\right] + \left[\frac{-10\times 10\times 10}{1\times 2}\right] + \left[\frac{-10\times 10\times 10}{\frac{1}{2}}\right]$$

$$= 1000(-3)$$

$$v_{13} = v_{31} = \left[\frac{-10\times 10\times 10}{1\times 2}\right] 2 + [0] + [0]$$

$$= 1000(-1)$$

$$v_{22} = \left[\frac{10\times 10\times 10}{1} + \frac{10\times 10\times 10}{1\times 3}\right] + \left[\frac{10\times 10\times 10}{1\times 3}\right] 2 + \left[\frac{10\times 10\times 10}{\frac{1}{2}}\right] 2$$

$$= 1000(6)$$

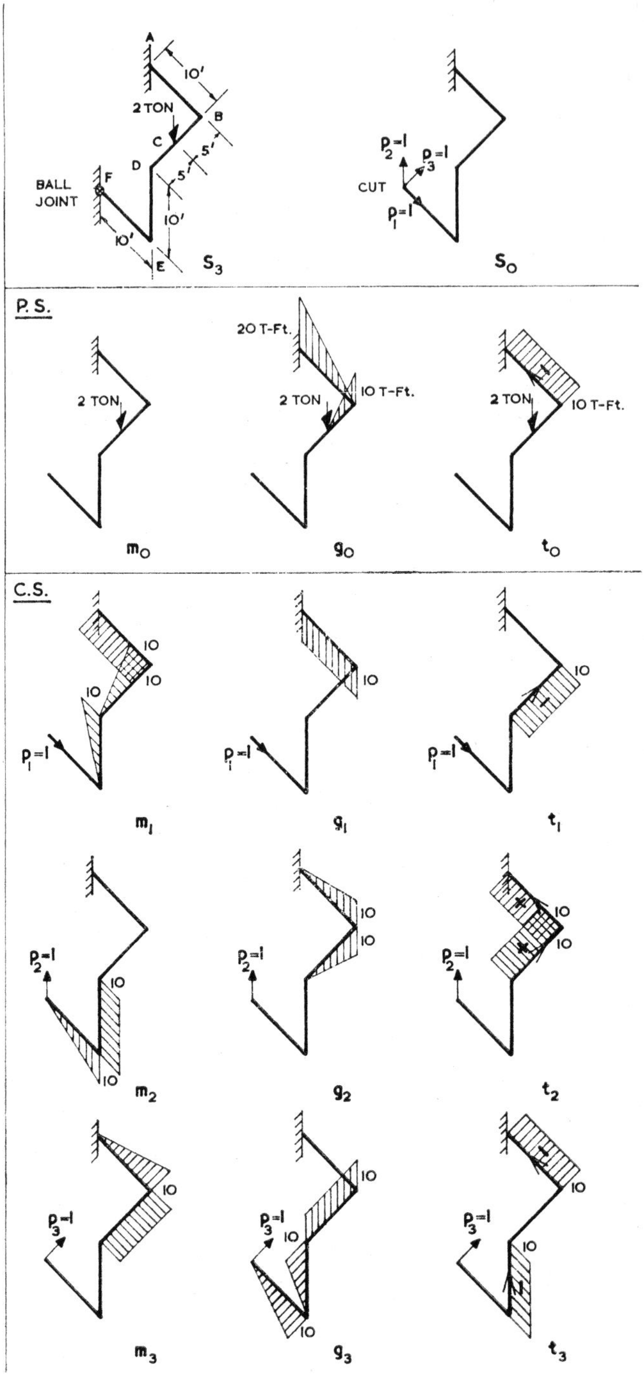

A
10'
2 TON
B
C
D
5'
5'
BALL JOINT
F
10'
10'
E
S3
P2=1
P3=1
CUT
P1=1
So
P. S.
2 TON
mo
20 T-Ft.
2 TON
10 T-Ft.
go
2 TON
10 T-Ft.
to
C.S.
10
10
10
P1=1
m1
10
P1=1
g1
10
P1=1
t1
P2=1
10
10
m2
10
10
P2=1
g2
10
10
P2=1
t2
10
P3=1
m3
10
P3=1
10
10
g3
10
P3=1
10
t3

FIG. 22.

$$v_{23}=v_{32}=[0]+\left[\frac{-10\times 10\times 10}{1\times 2}\right]+\left[\frac{-10\times 10\times 10}{½}\right]$$

$$=1000\left(-\tfrac{5}{2}\right)$$

$$v_{33}=\left[\frac{10\times 10\times 10}{1\times 3}+\frac{10\times 10\times 10}{1}\right]+\left[\frac{10\times 10\times 10}{1}+2\frac{(10\times 10\times 10)}{1\times 3}\right]+\left[\frac{10\times 10\times 10}{½}\right]2$$

$$=1000(7)$$

$$v_{10}=[0]+\left[-\frac{10\times 10\times 20}{1\times 2}\right]+[0]$$

$$=1000(-1)$$

$$v_{20}=[0]+\left[\frac{10\times 10\times 20}{1\times 6}-\frac{5}{1\times 3\times 2}(0+4\times 7{\cdot}5\times 5{\cdot}0+10\times 10)\right]$$

$$+\left[-\frac{10\times 10\times 10}{½}\right]$$

$$=1000(-1{\cdot}875)$$

$$v_{30}=[0]+\left[\frac{5\times 10\times 10}{1\times 2}\right]+\left[\frac{10\times 10\times 10}{½}\right]$$

$$=1000(2{\cdot}25)$$

The above flexibility coefficients give the relation

$$\boldsymbol{V}\cdot\boldsymbol{p}=-\boldsymbol{v}_0$$

$$1000\begin{bmatrix}\frac{14}{3} & -3 & -1\\ -3 & 6 & -\frac{5}{2}\\ -1 & -\frac{5}{2} & 7\end{bmatrix}\begin{bmatrix}p_1\\ p_2\\ p_3\end{bmatrix}=1000\begin{bmatrix}1{\cdot}000\\ 1{\cdot}875\\ -2{\cdot}250\end{bmatrix}$$

or

$$\boldsymbol{p}=-\boldsymbol{V}^{-1}\cdot\boldsymbol{v}_0$$

$$\begin{bmatrix}p_1\\ p_2\\ p_3\end{bmatrix}=\begin{bmatrix}0{\cdot}431\,589\,54 & 0{\cdot}283\,702\,21 & 0{\cdot}162\,977\,86\\ 0{\cdot}283\,702\,21 & 0{\cdot}382\,293\,76 & 0{\cdot}177\,062\,37\\ 0{\cdot}162\,977\,86 & 0{\cdot}177\,062\,37 & 0{\cdot}229\,376\,26\end{bmatrix}\begin{bmatrix}1{\cdot}000\\ 1{\cdot}875\\ -2{\cdot}250\end{bmatrix}$$

Hence

$$p_1=0{\cdot}596\,830\,99$$

$$p_2=0{\cdot}602\,112\,68$$

$$p_3=-0{\cdot}021\,199\,29$$

The stress resultants are obtained in a tabular form from the expression

$$\boldsymbol{x}_t=\boldsymbol{H}\cdot\boldsymbol{p}+\boldsymbol{x}_0$$

and the P.S. and C.S. diagrams

i.e.

$$m_t=m_1p_1+m_2p_2+m_3p_3+m_0$$

$$g_t=g_1p_1+g_2p_2+g_3p_3+g_0$$

$$t_t=t_1p_1+t_2p_2+t_3p_3+t_0$$

TOTAL MOMENT IN PLANE OF FRAME m_t (Tons – Ft.)

m_t =	m_1p_1 +	m_2p_2 +	m_3p_3 +	m_0 =	m_t
m_t^{AB}	(–10)(0·596 830 99)	0	0	0	– 5·968 309 9
$m_t^{BA=BC}$	(–10)(0·596 830 99)	0	(10)(–0·021 199 29)	0	– 6·180 302 8
m_t^{DC}	0	0	(10)(–0·021 199 29)	0	– 0·211 992 9
m_t^{DE}	(–10)(0·596 830 99)	(10)(0·602 112 68)	0	0	+ 0·052 816 9
$m_t^{ED=EF}$	0	(10)(0·602 112 68)	0	0	+ 6·021 126 8

TOTAL MOMENT NORMAL TO PLANE OF FRAME g_t (Tons – Ft.)

g_t =	g_1p_1 +	g_2p_2 +	g_3p_3 +	g_0 =	g_t
g_t^{AB}	(–10)(0·596 830 99)	0	0	+20·0	+14·031 690 1
g_t^{BA}	(–10)(0·596 830 99)	(10)(0·602 112 68)	0	0	+ 0·052 816 9
g_t^{BC}	0	(–10)(0·602 112 68)	(10)(–0·021 199 29)	+10·0	+ 3·766 880 3
g_t^{C}	0	(–5)(0·602 112 68)	(10)(–0·021 199 29)	0	– 3·222 556 3
g_t^{DC}	0	0	(10)(–0·021 199 29)	0	– 0·211 992 9
g_t^{DE}	0	0	(10)(–0·021 199 29)	0	– 0·211 992 9
g_t^{ED}	0	0	0	0	0
g_t^{EF}	0	0	(10)(–0·021 199 29)	0	– 0·211 992 9

TOTAL TORQUE t_t (Tons – Ft.)

t_t =	t_1p_1 +	t_2p_2 +	t_3p_3 +	t_0 =	t_t
$t_t^{AB=BA}$	0	(10)(0·602 112 68)	(–10)(–0·021 199 29)	–10·0	– 3·766 880 3
$t_t^{DB=BD}$	(–10)(0·596 830 99)	(10)(0·602 112 68)	0	0	+ 0·052 816 9
$t_t^{DE=ED}$	0	0	(–10)(–0·021 199 29)	0	+ 0·211 992 9
t_t^{EF}	0	0	0	0	0

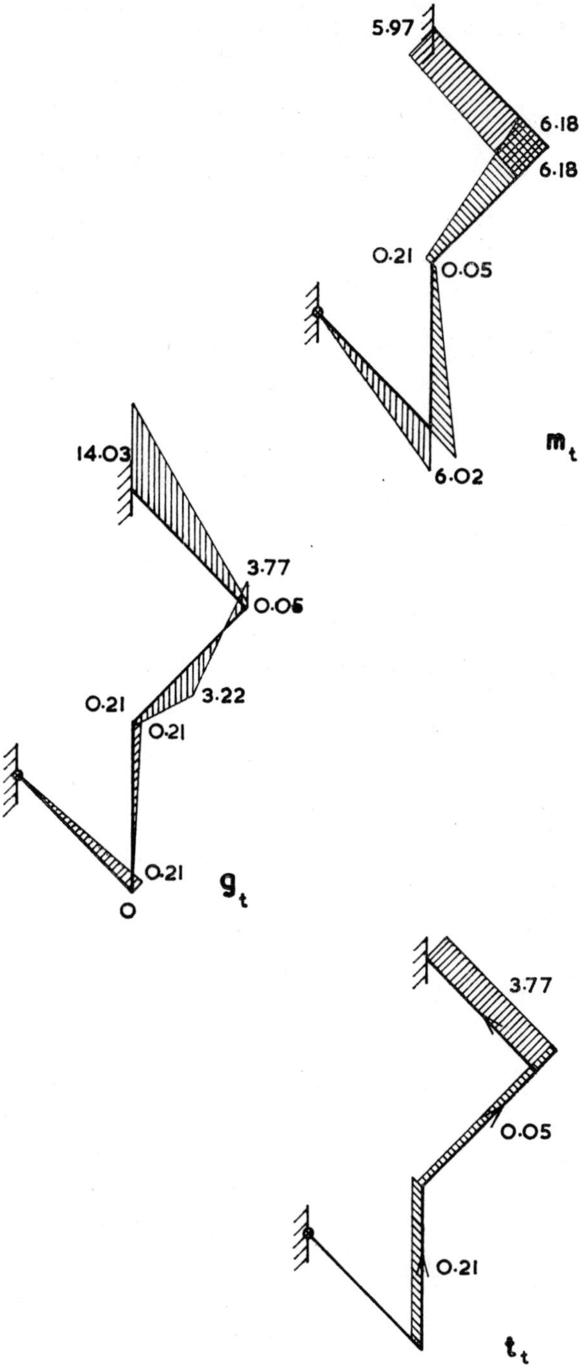

5·97
6·18
6·18
0·21
0·05
6·02
m_t
14·03
3·77
0·05
0·21
3·22
0·21
0·21
0
g_t
3·77
0·05
0·21
t_t

FIG. 23

The results are shown in Fig. 23.

Checks:

The following check equations were applied

$$\int \frac{m_1 m_t\, ds}{EI_1} + \int \frac{g_1 g_t\, ds}{EI_2} + \int \frac{t_1 t_t\, ds}{CJ} = 0 \qquad (= \delta v_1)$$

$$\int \frac{m_2 m_t\, ds}{EI_1} + \int \frac{g_2 g_t\, ds}{EI_2} + \int \frac{t_2 t_t\, ds}{CJ} = 0 \qquad (= \delta v_2)$$

$$\int \frac{m_3 m_t\, ds}{EI_1} + \int \frac{g_3 g_t\, ds}{EI_2} + \int \frac{t_3 t_t\, ds}{CJ} = 0 \qquad (= \delta v_3)$$

the values obtained were

$$\delta v_1 = +\ 0{\cdot}07253$$

$$\delta v_2 = +14{\cdot}24663$$

$$\delta v_3 = -\ 0{\cdot}43834$$

These are sufficiently small compared with the values of $\boldsymbol{v}_0$ for the solution to be considered correct.

Example C.

Draw the stress resultant diagrams for the tied cantilever shown in Fig. 24.

Moment of inertia of beam = 300 in^4

Cross Sectional Area of beam = 10 in^2

Cross Sectional Area of tie = 1·0 in^2

Solution

The structure is statically indeterminate to the first degree.

It will be made statically determinate by a cut at B in the tie rod AB.

Considering the bending and direct force energies only as significant in causing the deformations of the structure, we have from the P.S. and C.S. diagrams.

$$v_{11} = \int \frac{m_1 m_1\, ds}{EI} + \int \frac{n_1 n_1\, ds}{EA}$$

$$= \left[\frac{9{\cdot}6 \times 9{\cdot}6 \times 16}{300 \times 3E}\right] 1728 + \left[\frac{(-1{\cdot}0)(-1{\cdot}0)20}{1{\cdot}0E}\right] 12 + \left[\frac{0{\cdot}8 \times 0{\cdot}8 \times 16}{10E}\right] 12$$

$$= 1/E(2831{\cdot}16 + 240{\cdot}0 + 12{\cdot}29) = 3{,}083{\cdot}45/E$$

$$v_{10} = \frac{8 \times 12}{2 \times 3}[9{\cdot}6 \times 80 + 4 \times 40 \times 7{\cdot}2] \frac{144}{300E} = 14{,}745{\cdot}6/E$$

Now

$$v_{11} p_1 = -v_{10}$$

$$\therefore\ p_1 = -\frac{14{,}745{\cdot}6}{3{,}083{\cdot}45} = -4{\cdot}78$$

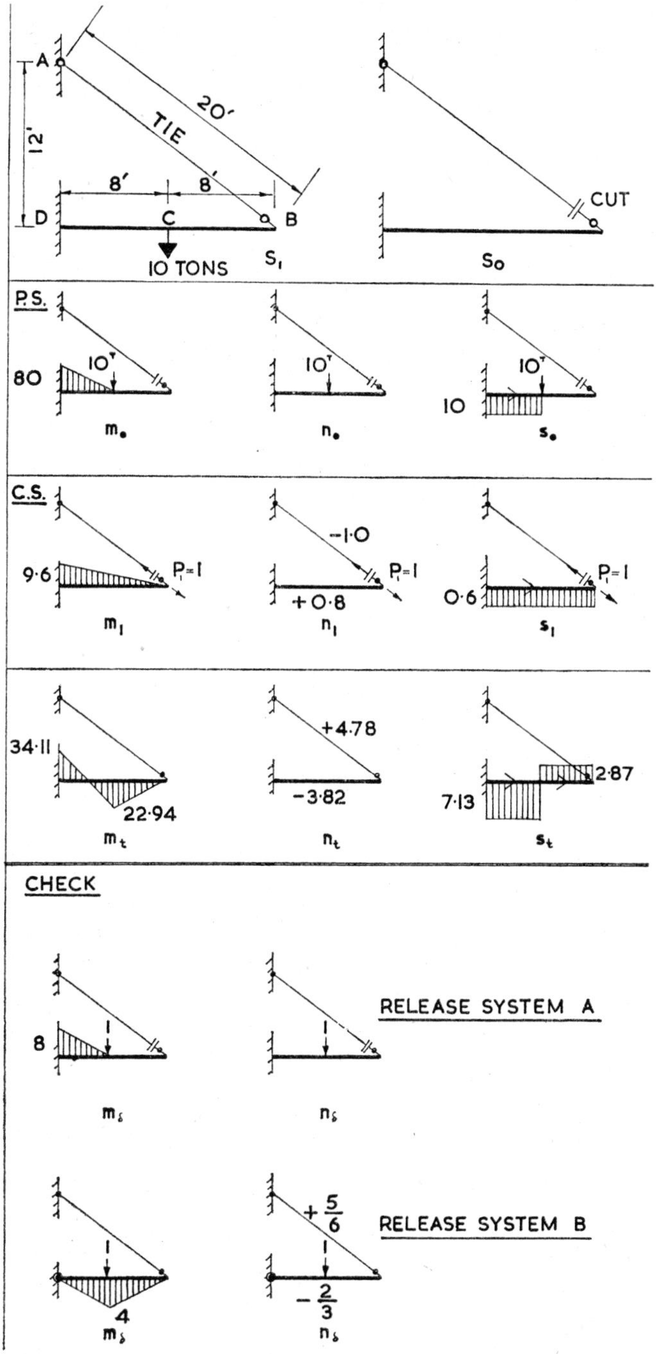
A
12'
TIE
20'
8'
8'
D
C
B
10 TONS
S₁
CUT
So
P.S.
80
10ᵀ
mₒ
10ᵀ
nₒ
10ᵀ
10
sₒ
C.S.
9·6
P₁=1
m₁
−1·0
P₁=1
+0·8
n₁
P₁=1
0·6
s₁
34·11
22·94
mₜ
+4·78
−3·82
nₜ
2·87
7·13
sₜ
CHECK
8
1
m₈
1
n₈
RELEASE SYSTEM A
1
4
m₈
+ 5/6
1
− 2/3
n₈
RELEASE SYSTEM B

FIG. 24.

If the axial force had been ignored then

$$p_1 = -\frac{14{,}745\cdot 6}{2831\cdot 15} = -5\cdot 22$$

The total stress resultants are obtained from

$$\boldsymbol{x}_t = \boldsymbol{Hp} + \boldsymbol{x}_0$$

i.e.

$$m_t = m_1 p_1 + m_0$$

$$n_t = n_1 p_1 + n_0$$

$$s_t = s_1 p_1 + s_0$$

Total Moments m_t

$$m_t^D = 9\cdot 6(-4\cdot 78) + 80 \qquad = +34\cdot 11 \text{ Tons-Ft}$$

$$m_t^C = 4\cdot 8(-4\cdot 78) + 0 \qquad = -22\cdot 94 \text{ Tons-Ft}$$

Total Axial Forces n_t

$$n_t^{AB} = -1\cdot 0(-4\cdot 78) + 0 \qquad = +4\cdot 78 \text{ Tons}$$

$$n_t^{BD} = 0\cdot 8(-4\cdot 78) + 0 \qquad = -3\cdot 82 \text{ Tons}$$

Total Shear Forces s_t

$$s_t^{BC} = 0\cdot 6(-4\cdot 78) + 0 \qquad = -2\cdot 87 \text{ Tons}$$

$$s_t^{CB} = 0\cdot 6(-4\cdot 78) + 0 \qquad = -2\cdot 87 \text{ Tons}$$

$$s_t^{CD} = 0\cdot 6(-4\cdot 78) + 10 \qquad = +7\cdot 13 \text{ Tons}$$

$$s_t^{D} = 0\cdot 6(-4\cdot 78) + 10 \qquad = +7\cdot 13 \text{ Tons.}$$

The final stress resultant diagrams m_t, n_t and s_t are shown in Fig. 24

Checks

Since the structure is one times statically indeterminate, the check will consist of calculating the deflection downwards at the point C using two release systems A and B as indicated in Fig. 24.

For Release System A

$$\delta_c = \int \frac{m_\delta m_t \, ds}{EI} + \int \frac{n_\delta n_t \, ds}{EA}$$

$$= \left[\tfrac{4}{3}(34\cdot 11 \times 8 + 4 \times 4 \times 5\cdot 585 + 0)\frac{1}{300E}\right]1728 + [0] = \frac{2780}{E}$$

For Release System B

$$\delta_c = \int \frac{m_\delta m_t \, ds}{EI} + \int \frac{n_\delta n_t \, ds}{EA}$$

$$= \left[\tfrac{4}{3}(-4 \times 5\cdot 585 \times 2 + 4 \times 22\cdot 94) + \left(4 \times 22\cdot 94 \times \tfrac{8}{3}\right)\right]\frac{1728}{300E}$$

$$+ \left[\tfrac{5}{6} \times \frac{4\cdot 78 \times 20}{E} + \tfrac{2}{3} \times \frac{3\cdot 82 \times 16}{10E}\right]12$$

$$= \frac{2780}{E}$$

This shows the answer to be correct.

Example D.

Compute the forces in all the members of the truss shown in Fig. 25. Assume that the cross sectional area A and elastic modulus E are constant for all members.

Solution

The truss is two times statically indeterminate and is released by cuts in members CF and CH.

From the P.S. and C.S. diagrams we have

$$v_{11}=\int\frac{n_1n_1\,ds}{EA}=\frac{1}{EA}[2\times0{\cdot}8\times0{\cdot}8\times10+2(-1)(-1)12{\cdot}5+2\times0{\cdot}6\times0{\cdot}6\times7{\cdot}5]$$

$$=\frac{43{\cdot}2}{EA}$$

It is clear from the C.S. diagrams that $v_{11}=v_{22}$ in this case.

$$v_{12}=v_{21}=\int\frac{n_1n_2\,ds}{EA}=\frac{1}{EA}[0{\cdot}8\times0{\cdot}8\times10]$$

$$=\frac{6{\cdot}4}{EA}$$

$$v_{10}=\int\frac{n_1n_0\,ds}{EA}=\frac{1}{EA}[16\times0{\cdot}8\times10+(-5)(-1)12{\cdot}5+(-6)0{\cdot}6\times7{\cdot}5$$

$$+9\times0{\cdot}6\times7{\cdot}5]=\frac{204}{EA}$$

$$v_{20}=\int\frac{n_2n_0\,ds}{EA}=\frac{1}{EA}[-6\times0{\cdot}6\times7{\cdot}5+5(-1)12{\cdot}5+3\times0{\cdot}6\times7{\cdot}5]$$

$$=-\frac{76}{EA}$$

The above flexibility coefficients give the relation

$$\boldsymbol{V}\cdot\boldsymbol{p}=-\boldsymbol{v}_0$$

$$\frac{1}{EA}\begin{bmatrix}43{\cdot}2 & 6{\cdot}4\\ 6{\cdot}4 & 43{\cdot}2\end{bmatrix}\begin{bmatrix}p_1\\ p_2\end{bmatrix}=\frac{1}{EA}\begin{bmatrix}-204\\ 76\end{bmatrix}$$

$$\therefore\quad \boldsymbol{p}=-\boldsymbol{V}^{-1}\cdot\boldsymbol{v}_0$$

$$\begin{bmatrix}p_1\\ p_2\end{bmatrix}=\frac{1}{1825{\cdot}28}\begin{bmatrix}43{\cdot}2 & -6{\cdot}4\\ -6{\cdot}4 & 43{\cdot}2\end{bmatrix}\begin{bmatrix}-204\\ 76\end{bmatrix}=\begin{bmatrix}-5{\cdot}09\\ 2{\cdot}51\end{bmatrix}$$

The total stress resultants are obtained in a tabular form below from the relation

$$\boldsymbol{x}_t=\boldsymbol{H}\cdot\boldsymbol{p}+\boldsymbol{x}_0$$

and the P.S. and C.S. diagrams

i.e. $$n_t=n_1p_1+n_2p_2+n_0$$

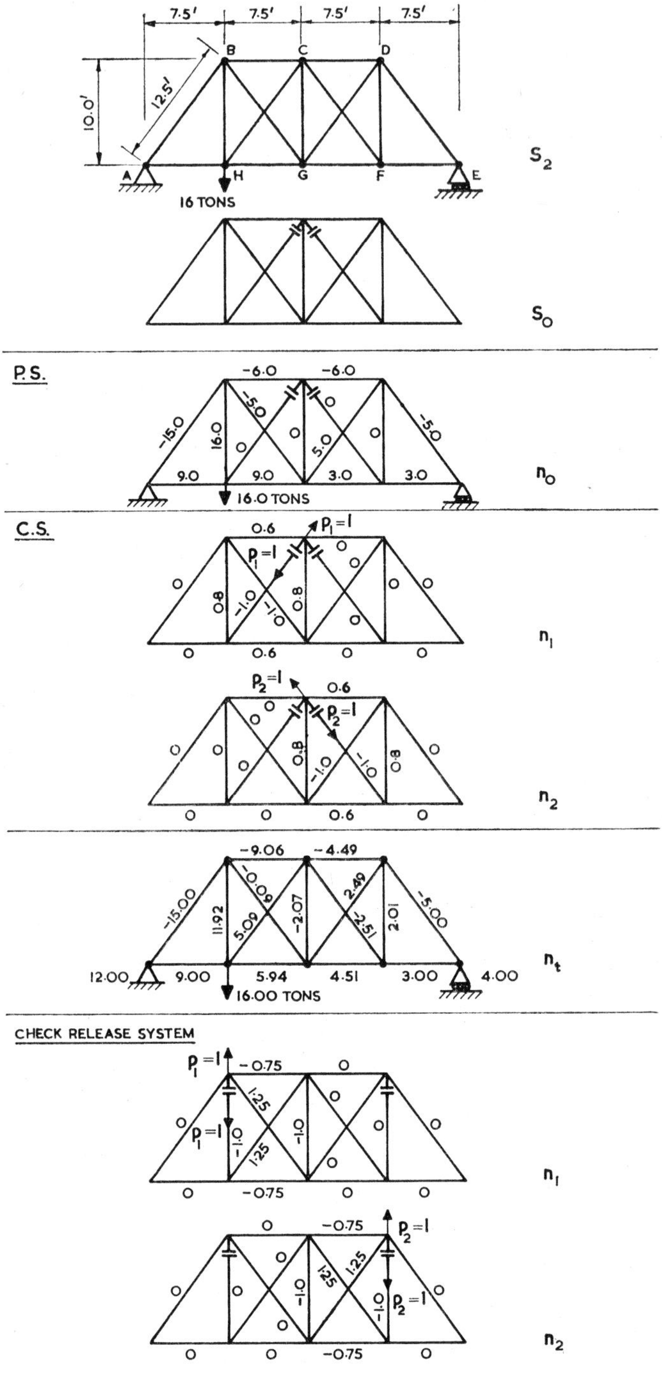
7.5′
7.5′
7.5′
7.5′
10.0′
12.5′
A
B
C
D
E
F
G
H
16 TONS
S_2
S_0
P.S.
−6.0
−6.0
−15.0
16.0
−5.0
5.0
−5.0
9.0
9.0
3.0
3.0
16.0 TONS
n_0
C.S.
$P_1=1$
$P_1=1$
0.6
0.8
−1.0
n_1
$P_2=1$
$P_2=1$
n_2
−9.06
−4.49
−15.00
11.92
−0.09
5.09
−2.07
2.49
−2.51
2.01
−5.00
12.00
9.00
5.94
4.51
3.00
4.00
16.00 TONS
n_t
CHECK RELEASE SYSTEM
−0.75
1.25
−1.0

FIG. 25.

MEMBER	n_1p_1	$+ \quad n_2p_2$	$+ \quad n_0$	$= \quad n_t$
AB	0	0	–15·00	–15·00
BC	0·6(–5·09)	0	– 6·00	– 9·06
CD	0	0·6(2·51)	– 6·00	– 4·49
DE	0	0	– 5·00	– 5·00
EF	0	0	3·00	+ 3·00
FG	0	0·6(2·51)	3·00	+ 4·51
GH	0·6(–5·09)	0	9·00	+ 5·94
HA	0	0	9·00	+ 9·00
HB	0·8(–5·09)	0	16·00	+11·92
GC	0·8(–5·09)	0·8(2·51)	0	– 2·07
FD	0	0·8(2·51)	0	+ 2·01
HC	–1·0(–5·09)	0	0	+ 5·09
BG	–1·0(–5·09)	0	– 5·00	– 0·09
CF	0	–1·0(2·51)	0	– 2·51
GD	0	–1·0(2·51)	5·00	+ 2·49

Checks

A different release system was used in the checking of the results obtained.

This consisted in cutting bars *BH* and *DF*.

1ST *Check*

$$\int \frac{n_1 n_t\, ds}{EA} = 0 \qquad (=\delta v_1)$$

$$= \frac{1}{EA}[(-0{\cdot}75)(-9{\cdot}06)7{\cdot}5 + (-0{\cdot}75)(5{\cdot}94)7{\cdot}5 + (-1{\cdot}0)(11{\cdot}92)10$$
$$+ (-1{\cdot}0)(-2{\cdot}07)10 + (1{\cdot}25)(-0{\cdot}09)12{\cdot}5 + (1{\cdot}25)(5{\cdot}09)12{\cdot}5]$$

$$= -\frac{2{\cdot}82}{EA} = \delta v_1$$

2nd *Check*

$$\int \frac{n_2 n_t\, ds}{EA} = 0 \qquad (=\delta v_2)$$

$$= \frac{1}{EA}[(-0{\cdot}75)(-4{\cdot}49)7{\cdot}5 + (-0{\cdot}75)(+4{\cdot}51)7{\cdot}5 + (-1{\cdot}0)(-2{\cdot}07)10$$
$$+ (-1{\cdot}0)(2{\cdot}01)10 + (1{\cdot}25)(-2{\cdot}51)12{\cdot}5 + (1{\cdot}25)(2{\cdot}49)12{\cdot}5]$$

$$= \frac{0{\cdot}175}{EA} = \delta v_2$$

The residuals δv_1 and δv_2 are sufficiently small compared with the values of $\boldsymbol{v}_0$ for the solution to taken as correct.

TEMPERATURE EFFECTS, LACK OF FIT AND SETTLEMENT OF SUPPORTS

Statically determinate structures are capable of accommodating temperature changes and settlement of supports without affecting the stress resultants in the structure.

Statically indeterminate structures, on the other hand, are generally restrained and any temperature changes, initial lack of fit of members, or the relative settlement of supports will induce stress resultants in the structure.

The effects of temperature, lack of fit and settlement of supports are of a similar nature to those produced by loading, except, of course, the causes are different.

Temperature Effects

In calculating the temperature effects in a structure, the following assumptions are made

(i) The value of the Elastic Modulus E remains constant over the range of temperatures.

(ii) The temperature gradient between the two surfaces of a member is linear.

Consider an element of length ds of a planar structural member subject to temperatures T_1, and T_2, as shown below in Fig. 26.

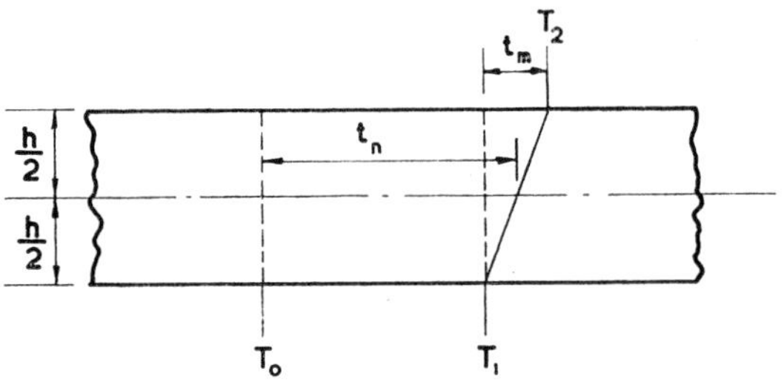

TEMPERATURE VARIATION IN A MEMBER

FIG. 26

T_0 = Temperature at which there is zero strain in the member

$$t_m = (T_1 - T_2)$$

$$t_n = \frac{(T_1 + T_2)}{2} - T_0$$

The corresponding unrestrained deformations of an element of length ds shown in Fig. 27 yield

$$(du_1)_{\text{temp}} = \frac{C}{h} \cdot t_m \cdot ds$$

$$(du_2)_{\text{temp}} = C \cdot t_n \cdot ds$$

where C = coefficient of thermal expansion, or in matrix form

$$\begin{bmatrix} du_1 \\ du_2 \end{bmatrix}_{\text{temp}} = \begin{bmatrix} \frac{C}{h} & 0 \\ 0 & C \end{bmatrix} \begin{bmatrix} t_m \\ t_n \end{bmatrix} ds$$

i.e. $$\boldsymbol{du}_{(\text{temp})} = \boldsymbol{F}_{(\text{temp})} \cdot \boldsymbol{t} \cdot ds$$

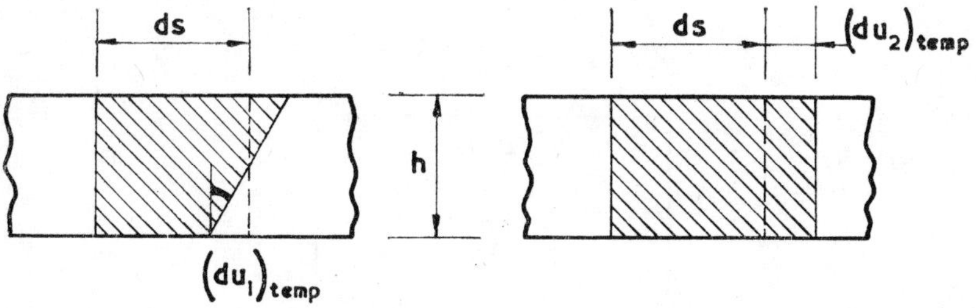

FIG. 27

The discontinuities produced in the structure at the releases (corresponding to the particular solution $\boldsymbol{v}_0$ in the case of loading) are given by the column matrix $\boldsymbol{v}_{(\text{temp})}$

From equations (3) and (5)

$$\boldsymbol{dv}_{(\text{temp})} = \boldsymbol{K} \cdot \boldsymbol{du}_{(\text{temp})}$$
$$= \boldsymbol{H}^{\text{T}} \cdot \boldsymbol{du}_{(\text{temp})}$$

$$\therefore \quad \boldsymbol{v}_{(\text{temp})} = \int \boldsymbol{H}^{\text{T}} \cdot \boldsymbol{du}_{(\text{temp})} = \int \boldsymbol{H}^{\text{T}} \cdot \boldsymbol{F}_{(\text{temp})} \cdot \boldsymbol{t} \cdot ds$$

The elements of the column matrix $\boldsymbol{v}_{(\text{temp})}$ are

$$v_{rt} = \int m_r \left(\frac{C}{h}\right) \cdot t_m \cdot ds + \int n_r (C) \cdot t_n \cdot ds$$

The Complementary Solution gives the matrix $\boldsymbol{V}$ as before. The equation of compatibility becomes

$$\underline{\boldsymbol{V} \cdot \boldsymbol{p}_{(\text{temp})} + \boldsymbol{v}_{(\text{temp})} = \boldsymbol{o}} \qquad (17)$$

and the total stress resultants are given by

$$\underline{\boldsymbol{x}_t = \boldsymbol{H} \cdot \boldsymbol{p}_{(\text{temp})}} \qquad (18)$$

since $\boldsymbol{x}_0 = \boldsymbol{o}$

Procedure

1. Calculate α_s
2. Make S_α statically determinate by providing suitable releases.
3. Draw the temperature diagrams (t_m, t_n).
4. Draw the C.S. diagrams (H^T).
5. Evaluate the elements of $\boldsymbol{V}$ and $\boldsymbol{v}_{(\text{temp})}$ from 3 and 4 above.
6. Compute $\boldsymbol{V}^{-1}$.
7. Obtain $\boldsymbol{p}_{(\text{temp})}$ from $\boldsymbol{p}_{(\text{temp})} = -\boldsymbol{V}^{-1} \cdot \boldsymbol{v}_{(\text{temp})}$.
8. Obtain $\boldsymbol{x}_t$ from $\boldsymbol{x}_t = \boldsymbol{H} \cdot \boldsymbol{p}_{(\text{temp})}$

Worked Example E

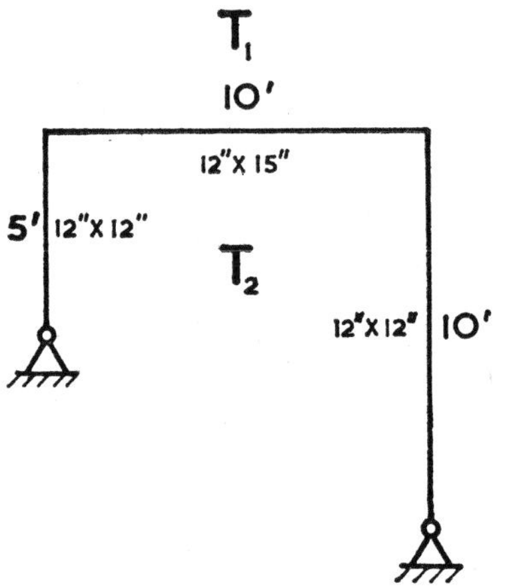

FIG. 28

Calculate the bending moments and axial forces in the concrete portal frame shown in Fig. 28 caused by the following temperature changes

$$T_1 = 100\,°\text{F} \quad T_2 = 80\,°\text{F} \quad T_0 = 50\,°\text{F}$$

$E = 2 \times 10^6$ lbs/sq. in and the linear coefficient of thermal expansion $C = \frac{1}{180{,}000} / °\text{F}$

Solution

The structure is made statically determinate by providing a roller at the lower support

$$t_m = 100 - 80 = 20\,°\text{F}$$

$$t_n = \frac{(100 + 80)}{2} - 50 = 40\,°\text{F}$$

$$I_{\text{COLUMN}} = \frac{1 \times 1^3}{12} = 0{\cdot}0833 \text{ ft}^4.$$

$$I_{\text{BEAM}} = \frac{1 \times (1{\cdot}25)^3}{12} = 0{\cdot}1628 \text{ ft}^4.$$

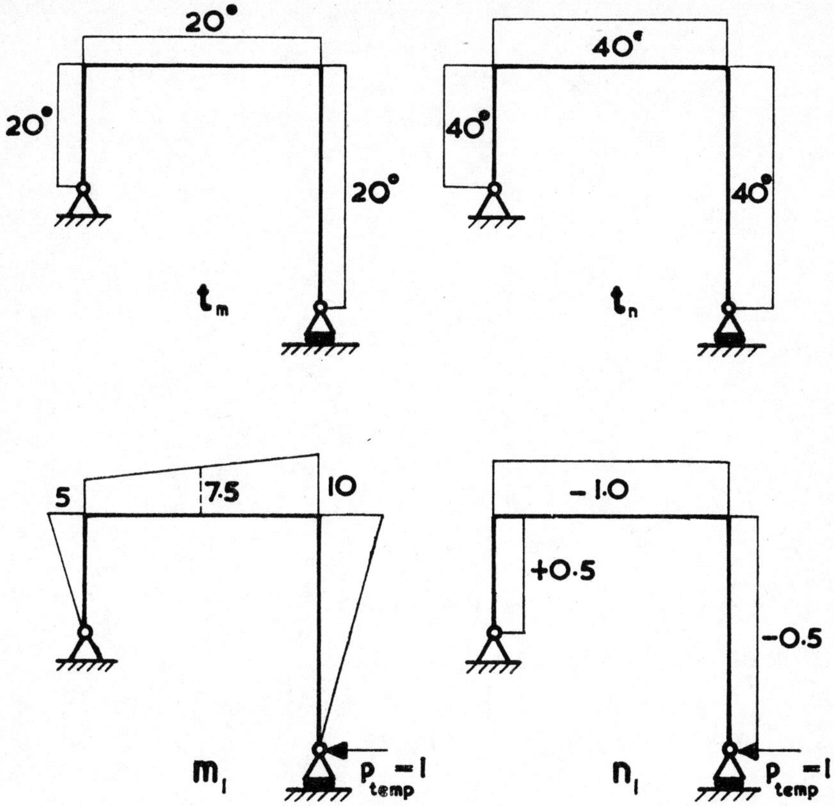

FIG. 29

$$v_{11} = \int \frac{m_1 m_1\, ds}{EI} + \int \frac{n_1 n_1\, ds}{EA}$$

$$= \frac{1}{E}\left[\frac{5\times 5\times 5}{3\times 0{\cdot}0833} + \frac{10}{6\times 0{\cdot}1628}(5\times 5 + 4\times 7{\cdot}5\times 7{\cdot}5 + 10\times 10) + \frac{10\times 10\times 10}{3\times 0{\cdot}0833}\right]$$

$$+ \frac{1}{E}\left[\frac{0{\cdot}5\times 0{\cdot}5\times 5}{1{\cdot}0} + \frac{(-1{\cdot}0)(-1{\cdot}0)10}{1{\cdot}25} + \frac{(-0{\cdot}5)(-0{\cdot}5)10}{1{\cdot}0}\right]$$

$$= \frac{1}{E}[500 + 3{,}570 + 4{,}000 + 1{\cdot}25 + 8{\cdot}0 + 2{\cdot}5] = \frac{8081{\cdot}75}{E}$$

NOTE. The direct energy terms are so small that they could have been omitted from the calculations in this particular case.

$$v_{(\text{temp})} = \int m_1\left(\frac{C}{h}\right)\cdot t_m\, ds + \int n_1(C) t_n \cdot ds$$

$$= \left[\frac{5\times 5\times 20}{2} + \frac{7{\cdot}5\times 10\times 20}{1{\cdot}25} + \frac{10\times 10\times 20}{2}\right]\frac{1}{180{,}000}$$

$$+ [0{\cdot}5\times 5\times 40 - 1{\cdot}0\times 10\times 40 - 0{\cdot}5\times 10\times 40]\frac{1}{180{,}000}$$

$$= [250 + 1{,}200 + 1{,}000 + 100 - 400 - 200]\frac{1}{180{,}000} = \frac{1{,}950}{180{,}000}$$

$$\therefore \quad p_{(\text{temp})} = -\frac{1}{v_{11}} \cdot v_{(\text{temp})} = \frac{1{,}950 \times 2 \times 144 \times 10^6}{180{,}000 \times 8081\cdot 75} = -387\cdot 0$$

$$\boldsymbol{x}_t = \boldsymbol{H} \cdot p_{(\text{temp})}$$

i.e.

$$m_t = m_1 p_{(\text{temp})}$$

$$n_t = n_1 p_{(\text{temp})}$$

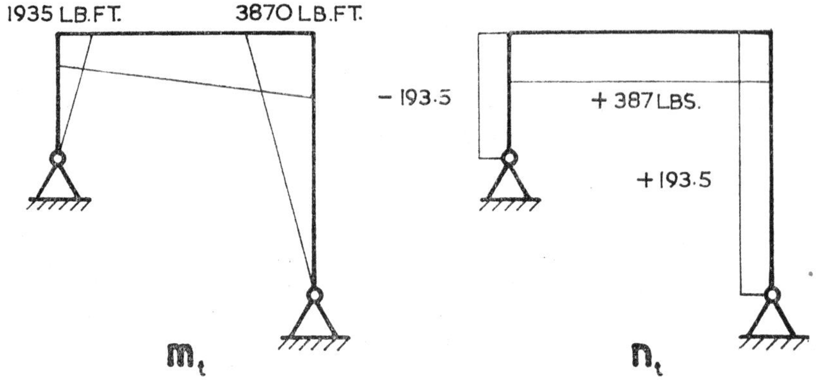

FIG. 30

Lack of Fit and Settlement of Supports

The stress resultants produced in a statically indeterminate structure as a result of lack of fit of prefabricated members or the forced yield of the supports can be evaluated in one of two ways:

(A) By putting the releases at the points of yield or lack of fit,

or

(B) By putting the releases at other points than those of yield or initial lack of fit.

In approach *A*, let $\boldsymbol{v}_t = \boldsymbol{\delta}$ be the set of yields or lack of fit displacements imposed at the releases, then, with no external loads acting on the structure, equation (11) $\boldsymbol{V} \cdot \boldsymbol{p} + \boldsymbol{v}_0 = \boldsymbol{v}_t$ becomes

$$\boldsymbol{V} \cdot \boldsymbol{p} + \boldsymbol{o} = \boldsymbol{\delta} \qquad \underline{\boldsymbol{V} \cdot \boldsymbol{p} = \boldsymbol{\delta}} \tag{19}$$

In approach *B*, let the imposed yields or lack of fit displacements $\boldsymbol{\delta}$ cause movements $\boldsymbol{v}_\delta$ at the releases, then, with no loads acting, equation (11) becomes

$$\boldsymbol{V} \cdot \boldsymbol{p} + \boldsymbol{v}_\delta + \boldsymbol{o} = \boldsymbol{o} \qquad \underline{\boldsymbol{V} \cdot \boldsymbol{p} = -\boldsymbol{v}_\delta} \tag{20}$$

The relation between the yields or lack of fit displacements and the movements $\boldsymbol{v}_\delta$ at the releases is, from equations (3) and (5)

$$\boldsymbol{v}_\delta = \boldsymbol{K} \cdot \boldsymbol{\delta} \qquad \text{(similar to } \boldsymbol{dV} = \boldsymbol{K} \cdot \boldsymbol{dU})$$

or

$$\underset{(\alpha\times 1)\,(\alpha\times\beta)\,(\beta\times 1)}{\underline{\boldsymbol{v}_\delta = \boldsymbol{H}^{\mathrm{T}} \cdot \boldsymbol{\delta}}} \tag{21}$$

where α is the statical indeterminacy of the structure, and β is the number of yields or lack of fit displacements in the structure.

For a single type of yield of support or lack of fit $\boldsymbol{\delta}$, equation (21) becomes

$$\underset{(\alpha\times 1)}{\boldsymbol{v}_{\delta}} = \underset{(\alpha\times 1)}{\boldsymbol{h}^{\mathrm{T}}} \cdot \underset{(1\times 1)}{\boldsymbol{\delta}}$$

where $\boldsymbol{h}^{\mathrm{T}}$ represents the appropriate column of the matrix $\boldsymbol{H}^{\mathrm{T}}$.

The column matrix $\boldsymbol{h}^{\mathrm{T}}$ depends on the type of yield of support or lack of fit.

For a *rotational* yield of a support or lack of fit at a point, $\boldsymbol{h}^{\mathrm{T}}$ will contain the *moment* effects at that point due to unit bi-actions at the releases.

i.e. $$\boldsymbol{h}^{\mathrm{T}} = \{m_1 \quad m_2 \ldots\ldots m_{\alpha}\}$$

For a *transverse* yield of a support or lack of fit at a point, $\boldsymbol{h}^{\mathrm{T}}$ will contain the *shear* effects at that point due to unit bi-actions at the releases.

i.e. $$\boldsymbol{h}^{\mathrm{T}} = \{s_1 \quad s_2 \ldots\ldots s_{\alpha}\}$$

For a *longitudinal* (axial) yield of support or lack of fit at a point, $\boldsymbol{h}^{\mathrm{T}}$ will contain the *axial force* effects at that point due to unit bi-actions at the releases.

i.e. $$\boldsymbol{h}^{\mathrm{T}} = \{n_1 \quad n_2 \ldots\ldots n_{\alpha}\} \quad \text{etc.}$$

The stress resultants produced are then calculated from the equation

$$\underline{\boldsymbol{x}_t = \boldsymbol{H} \cdot \boldsymbol{p}}$$

since $\boldsymbol{x}_0 = 0$ in the case of no applied loads.
The following examples illustrate both approaches A and B.

Worked Example F—Yield of Supports

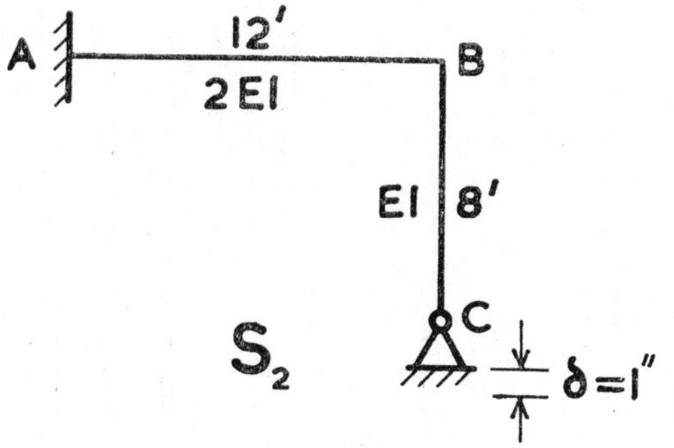

FIG. 31

Draw the bending moment diagram for the frame shown in Fig. 31 produced by a forced downward yield $\delta = 1''$ at C.
Take $EI = 10{,}000$ T-Ft²

Solution A, by providing release at the point of yield.

The structure is made statically determinate by inserting a cut at C. Consider bending energy as the only significant term.

$$v_{rs} = \int \frac{m_r m_s \, ds}{EI}$$

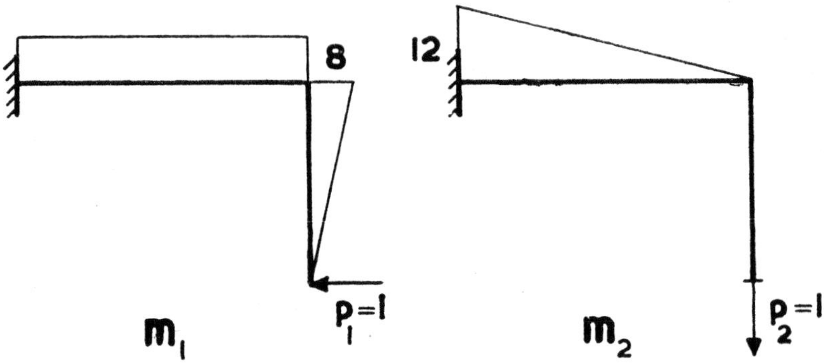

FIG. 32(a)

$$v_{11} = \frac{1}{E}\left[\frac{8\times 8\times 12}{2I} + \frac{8\times 8\times 8}{3I}\right] = \frac{32\times 52}{3EI}$$

$$v_{12} = v_{21} = \frac{1}{E}\left[12\times\frac{12}{2}\times\frac{8}{2I}\right] = \frac{288}{EI}$$

$$v_{22} = \frac{1}{E}\left[12\times\frac{12}{3}\times\frac{12}{2I}\right] = \frac{288}{EI}$$

Equation (19), $\quad \boldsymbol{V}\cdot\boldsymbol{p} = \boldsymbol{\delta}$

$$\frac{1}{EI}\begin{bmatrix} \frac{32\times 52}{3} & 288 \\ 288 & 288 \end{bmatrix}\begin{bmatrix} p_1 \\ p_2 \end{bmatrix} = \begin{bmatrix} 0 \\ \frac{1}{12} \end{bmatrix}$$

$$|V| = \frac{32\times 52\times 288}{3} - 288^2 = 288\times 266{\cdot}67$$

$$\therefore \begin{bmatrix} p_1 \\ p_2 \end{bmatrix} = \frac{10{,}000}{288\times 266{\cdot}67}\begin{bmatrix} 288 & -288 \\ -288 & \frac{32\times 52}{3} \end{bmatrix}\begin{bmatrix} 0 \\ \frac{1}{12} \end{bmatrix} = \begin{bmatrix} -3{\cdot}12 \\ 6{\cdot}02 \end{bmatrix}$$

From $\quad \boldsymbol{x}_t = \boldsymbol{H}\cdot\boldsymbol{p}$

$$m_t^{AB} = m_1 p_1 + m_2 p_2$$
$$= 8(-3{\cdot}12) + 12(6{\cdot}02)$$
$$= +47{\cdot}3\ T\text{-}Ft$$

$$m_t^{B} = m_1 p_1 + m_2 p_2$$
$$= 8(-3{\cdot}12) + 0$$
$$= -25{\cdot}0\ T\text{-}Ft$$

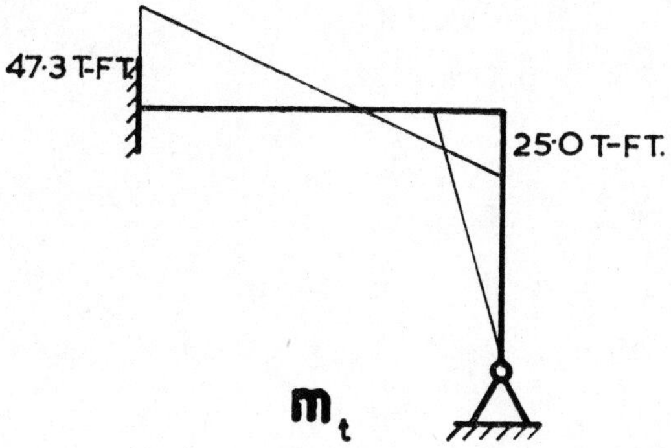

FIG. 32(b)

Solution B, by not providing releases at point of yield.

The structure is made statically determinate by providing a hinge and roller at A.

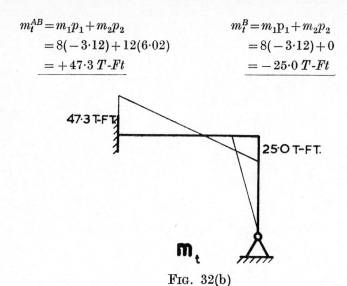

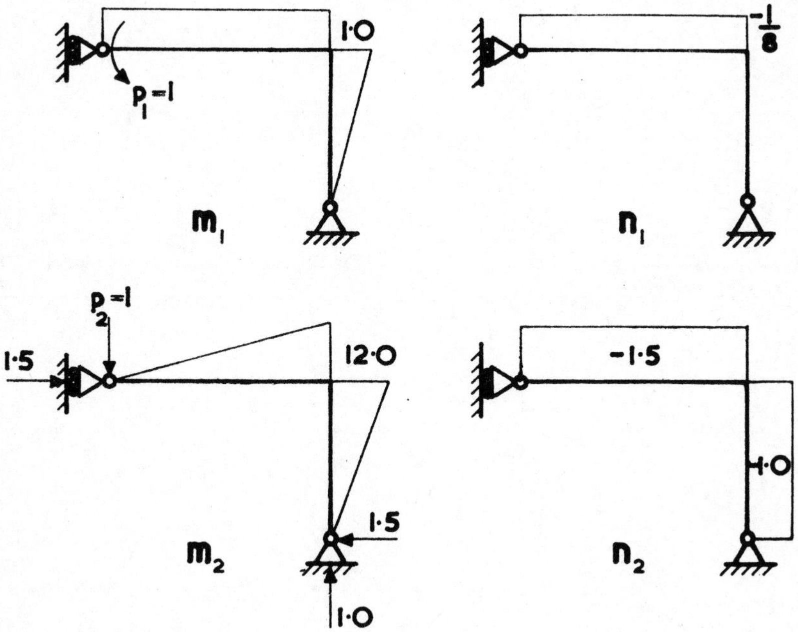

FIG. 33

In order to obtain the transformation matrix $\boldsymbol{h}^{\mathrm{T}}$ it is necessary to draw the axial force diagrams in addition to those for the bending moments.

$$\boldsymbol{V}\cdot\boldsymbol{p}=-\boldsymbol{v}_{\delta} \qquad \boldsymbol{v}_{\delta}=\boldsymbol{h}^{\mathrm{T}}\cdot\boldsymbol{\delta}$$
$$=-\boldsymbol{h}^{\mathrm{T}}\cdot\boldsymbol{\delta}$$

$$v_{11}=\frac{1}{E}\left[\frac{1\times1\times12}{2I}+\frac{1\times1\times8}{3I}\right]=\frac{26}{3EI}$$

$$v_{12}=v_{21}=\frac{1}{E}\left[\frac{1\times12\times12}{2\times2I}+\frac{1\times12\times8}{3I}\right]=\frac{68}{EI}$$

$$v_{22}=\frac{1}{E}\left[\frac{12\times12\times12}{2I\times3}+\frac{12\times12\times8}{I\times3}\right]=\frac{672}{EI}$$

$$\frac{1}{EI}\begin{bmatrix}\frac{26}{3} & 68\\ 68 & 672\end{bmatrix}\begin{bmatrix}p_1\\ p_2\end{bmatrix}=\begin{bmatrix}0\\ -1\end{bmatrix}\left(-\frac{1}{12}\right)$$

$$|V|=\frac{26}{3}\times672-68^2=1{,}200$$

$$\begin{bmatrix}p_1\\ p_2\end{bmatrix}=\frac{10{,}000}{1{,}200\times12}\begin{bmatrix}672 & -68\\ -68 & \frac{26}{3}\end{bmatrix}\begin{bmatrix}0\\ -1\end{bmatrix}=\begin{bmatrix}47{\cdot}3\\ -6{\cdot}02\end{bmatrix}$$

$$m_t^{AB}=m_1p_1+m_2p_2$$
$$=1(47{\cdot}3)+0=\underline{47{\cdot}3\ T\text{-ft.}}$$

$$m_t^{B}=m_1p_1+m_2p_2$$
$$=1(47{\cdot}3)+12(-6{\cdot}02)=\underline{-25{\cdot}0\ T\text{-ft.}}$$

The final Bending Moment diagram is shown in Fig. 32(b)

Worked Example G—Initial Lack of Fit.

Calculate the force in the member *BD* of the truss shown in Fig. 34 assuming that this member was prefabricated $\frac{1}{20}$ in. too short. $E=10{,}000$ Tons/sq · in.

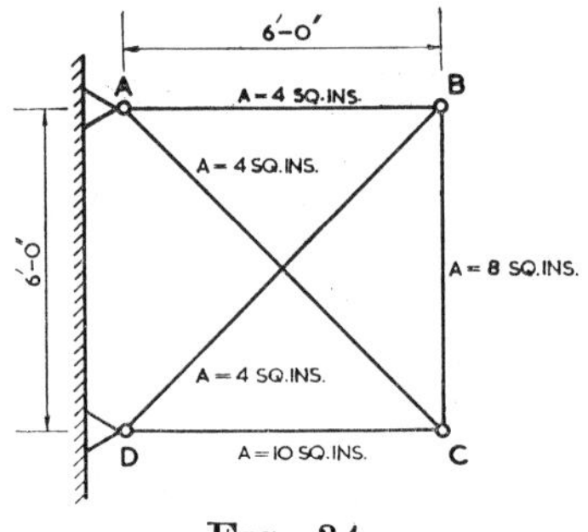

FIG. 34

Solution A

The truss is one times statically indeterminate and will be released by inserting a cut in member BD.

$$v_{11}=\left(\frac{1}{\sqrt{2}}\times\frac{1}{\sqrt{2}}\ \frac{6}{E}\right)\left(\frac{1}{4}+\frac{1}{8}+\frac{1}{10}\right)12+2\left(\frac{1\times1\times6\sqrt{2}}{4E}\right)12$$
$$=\frac{68\cdot1}{E}$$

$$v_{11}\cdot p_1=\delta$$

$$\therefore\quad p_1=\frac{\delta}{v_{11}}=-\frac{1}{20}\cdot\frac{10{,}000}{68\cdot1}=-7\cdot34$$

$\therefore$ Force in member $BD=p_1n_1=-7\cdot34(-1)=\underline{7\cdot34\text{ Tons}}$

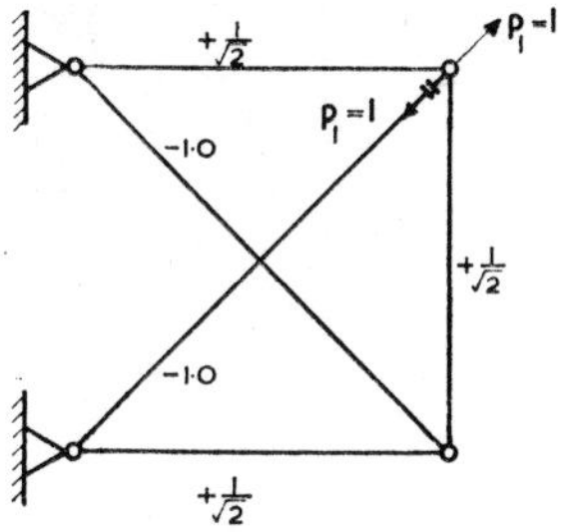

FIG. 35

Solution B

For this case the truss is released by inserting a cut in member BC.

$$\boldsymbol{v}_\delta=\boldsymbol{h}^{\mathrm{T}}\cdot\boldsymbol{u}$$
$$=\sqrt{2}\cdot\left(-\frac{1}{20}\right)=-\frac{\sqrt{2}}{20}$$
$$v_{11}p_1+v_\delta=0$$

$$v_{11}=1\times1\times\frac{6}{E}\left(\frac{1}{4}+\frac{1}{8}+\frac{1}{10}\right)12+2\left(\frac{\sqrt{2}\times\sqrt{2}\times6\sqrt{2}}{4E}\right)12$$
$$=\frac{72}{E}\times1\cdot89$$

$$\therefore\ p_1=-\frac{v_\delta}{v_{11}}=-\left(-\frac{\sqrt{2}}{20}\right)\frac{10{,}000}{72\times1\cdot89}=+5\cdot2$$

$\therefore$ Force in member $BD=p_1n_1=5\cdot2(\sqrt{2})=\underline{7\cdot34\text{ Tons}}$

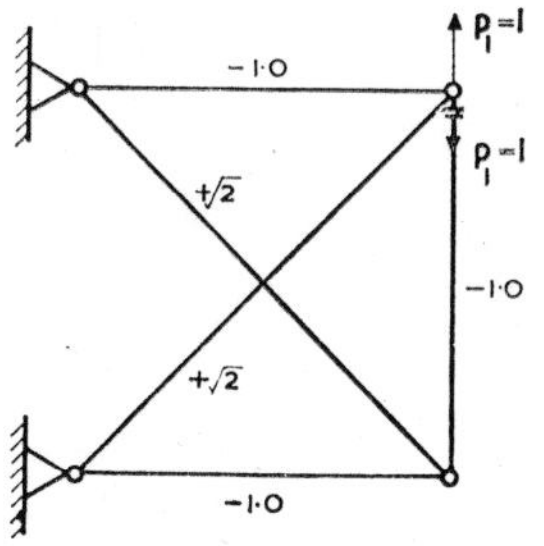

FIG. 36

THE STIFFNESS METHOD

Introduction

One of the most important features of the Stiffness Method is the way in which the elastic properties of the members are introduced into the analysis before the conditions of equilibrium and compatibility are considered.

Firstly, equations are developed relating the member end-loads to the end-displacements for each member. In this stage there is no need to worry about the complexity of the structure as a whole. Each node is rigidly fixed and therefore displacements cannot be transmitted from one member to another.

In the second stage, in which the conditions of equilibrium and compatibility at each node are introduced, only the way in which the members are linked need be considered.

The method is explained by following through the analysis of a plane frame in detail. The theory is developed including all three stress-resultants, i.e. bending moments, shear and axial forces. This approach is adopted since the theory is then applicable to any plane structure whether it is rigid or pin-jointed.

In practice the chosen example would be analysed using the bending terms only.

The analysis of a Rigid Plane Frame

Let us consider the plane frame shown in Fig. 37 in which all the joints are rigid

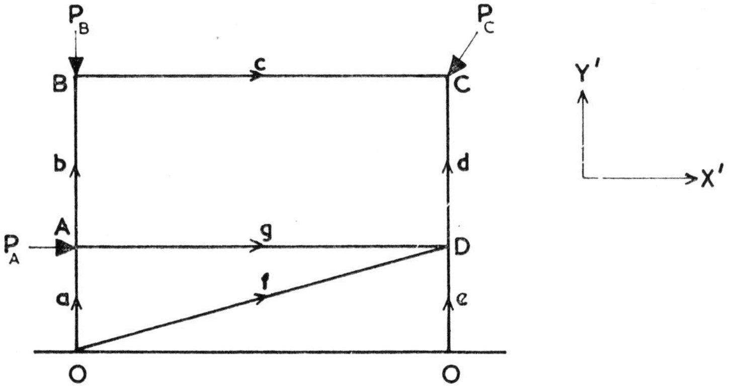

FIG. 37

For simplicity all the members will be regarded as straight and of uniform section. The material of which the members are made is assumed to obey Hooke's Law and the modulus of elasticity is taken as constant.

The member *f*, although clearly unnecessary, is added so that the structure contains a diagonal member. The reason for this will become clear later.

The properties of each member must be considered first. Since we are studying a plane frame only three stress resultants need be considered. The sign convention for each member is shown in Fig. 38.

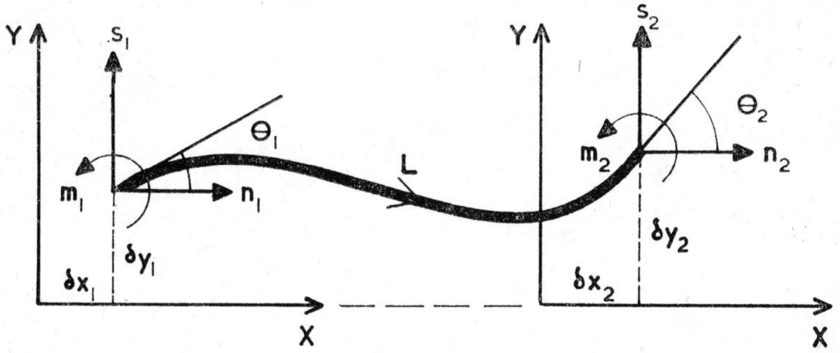

FIG. 38

In Fig. 38, *m*, *s*, and *n* are the moment, shear force and axial force respectively, at each end of the member. Θ, δy and δx are the corresponding deflexions. The subscripts 1 and 2 relate these to a particular end of the member. Each member is given a direction by means of an arrow as shown in Figs. 37 and 38. End 1 is always at the tail of the arrow, and end 2 at the head.

Thus, in Fig. 37, end 1 of member *g* is at *A* and end 2 at *D*. Similarly, for member *d*, end 1 is at *D* and end 2 at *C*.

The analysis will be restricted to linear small-deflexion theory and thus changes in the distance between the ends of a member due to curvature will be ignored.

The equations giving the end-loads in terms of end-displacements must now be set up. The only displacement giving rise to axial force is δx and therefore

$$n_1 = -n_2 = \frac{EA}{L}(\delta x_1 - \delta x_2) \qquad (22)$$

Referring to Fig. 38, the slope-deflection equations give

$$m_1 = \frac{6EI}{L^2}(\delta y_1 - \delta y_2)$$

and

$$m_2 = \frac{6EI}{L^2}(\delta y_1 - \delta y_2)$$

If a rotation Θ_1 is imposed on end 1, while end 2 is fixed, we have

$$m_1=\frac{4EI\Theta_1}{L} \quad \text{and} \quad m_2=\frac{2EI\Theta_1}{L}$$

Similarly, if a rotation Θ_2 is applied at end 2, and end 1 is fixed, then

$$m_1=\frac{2EI\Theta_2}{L} \quad \text{and} \quad m_2=\frac{4EI\Theta_2}{L}$$

Thus the total moments induced by the end-displacements are

$$m_1=\frac{6EI}{L^2}\delta y_1+\frac{4EI}{L}\Theta_1-\frac{6EI}{L^2}\delta y_2+\frac{2EI}{L}\Theta_2$$

$$m_2=\frac{6EI}{L^2}\delta y_1+\frac{2EI}{L}\Theta_1-\frac{6EI}{L^2}\delta y_2+\frac{4EI}{L}\Theta_2 \tag{23}$$

Taking moments about one end of the member gives

$$m_1+m_2=s_1L=-s_2L$$

From this and equations (23) we have

$$\begin{aligned} s_1&=-s_2\\ &=\left(\frac{12EI}{L^3}\right)\delta y_1+\left(\frac{6EI}{L^2}\right)\Theta_1-\left(\frac{12EI}{L^3}\right)\delta y_2+\left(\frac{6EI}{L^2}\right)\Theta_2 \end{aligned} \tag{24}$$

Combining equations (22), (23) and (24) and writing them in matrix form we obtain

$$\begin{bmatrix} m_1\\ s_1\\ n_1 \end{bmatrix}=\begin{bmatrix} \frac{4EI}{L} & \frac{6EI}{L^2} & 0\\ \frac{6EI}{L^2} & \frac{12EI}{L^3} & 0\\ 0 & 0 & \frac{EA}{L} \end{bmatrix}\begin{bmatrix} \Theta_1\\ \delta y_1\\ \delta x_1 \end{bmatrix}+\begin{bmatrix} \frac{2EI}{L} & -\frac{6EI}{L^2} & 0\\ \frac{6EI}{L^2} & -\frac{12EI}{L^3} & 0\\ 0 & 0 & -\frac{EA}{L} \end{bmatrix}\begin{bmatrix} \Theta_2\\ \delta y_2\\ \delta x_2 \end{bmatrix}$$

and

$$\begin{bmatrix} m_2\\ s_2\\ n_2 \end{bmatrix}=\begin{bmatrix} \frac{2EI}{L} & \frac{6EI}{L^2} & 0\\ -\frac{6EI}{L^2} & -\frac{12EI}{L^3} & 0\\ 0 & 0 & -\frac{EA}{L} \end{bmatrix}\begin{bmatrix} \Theta_1\\ \delta y_1\\ \delta x_1 \end{bmatrix}+\begin{bmatrix} \frac{4EI}{L} & -\frac{6EI}{L^2} & 0\\ -\frac{6EI}{L^2} & \frac{12EI}{L^3} & 0\\ 0 & 0 & \frac{EA}{L} \end{bmatrix}\begin{bmatrix} \Theta_2\\ \delta y_2\\ \delta x_2 \end{bmatrix}$$

Expressing these in a more compact form we have

$$\begin{aligned} \boldsymbol{x}_1&=\boldsymbol{S}_{11}\boldsymbol{d}_1+\boldsymbol{S}_{12}\boldsymbol{d}_2\\ \boldsymbol{x}_2&=\boldsymbol{S}_{21}\boldsymbol{d}_1+\boldsymbol{S}_{22}\boldsymbol{d}_2 \end{aligned} \tag{25}$$

Having set up these equations with reference to the co-ordinate system for each individual member, it is necessary to convert them to a common co-ordinate system. This co-ordinate system is called the structure system and is shown as X' and Y' in Fig. 37.

The relationship between the two systems is illustrated in Fig. 39. Throughout the text the primes (i.e. S') indicates that the term is related to the structure coordinates.

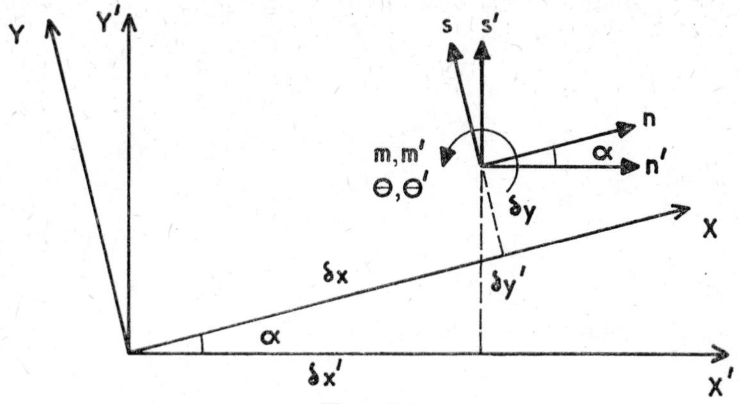

FIG. 39

The relationship between the stress resultants in the two systems is

$$\begin{bmatrix} m' \\ s' \\ n' \end{bmatrix} = \begin{bmatrix} 1 & 0 & 0 \\ 0 & \text{Cos } \alpha & \text{Sin } \alpha \\ 0 & -\text{Sin } \alpha & \text{Cos } \alpha \end{bmatrix} \begin{bmatrix} m \\ s \\ n \end{bmatrix}$$

where α is the angle between the member and the X' axis. Moments are unchanged by a rotation of the axes.

Similarly for the displacements

$$\begin{bmatrix} \Theta' \\ \delta y' \\ \delta x' \end{bmatrix} = \begin{bmatrix} 1 & 0 & 0 \\ 0 & \text{Cos } \alpha & \text{Sin } \alpha \\ 0 & -\text{Sin } \alpha & \text{Cos } \alpha \end{bmatrix} \begin{bmatrix} \Theta \\ \delta y \\ \delta x \end{bmatrix}$$

in matrix form we may write $\boldsymbol{x}' = \boldsymbol{T} \cdot \boldsymbol{x}$

$$\boldsymbol{d}' = \boldsymbol{T} \cdot \boldsymbol{d} \tag{26}$$

$\boldsymbol{T}$ is a square matrix with an inverse equal to its transpose. This may be easily shown by the matrix multiplication

$$\boldsymbol{T} \cdot \boldsymbol{T}^{\mathrm{T}} = \boldsymbol{I}$$

Therefore

$$\boldsymbol{x} = \boldsymbol{T}^{\mathrm{T}} \cdot \boldsymbol{x}'$$

and

$$\boldsymbol{d} = \boldsymbol{T}^{\mathrm{T}} \cdot \boldsymbol{d}'$$

By means of these equations the $\boldsymbol{x}$ and $\boldsymbol{d}$ terms in equations (25) may be eliminated.

$$\boldsymbol{T}^{\mathrm{T}} \cdot \boldsymbol{x}_1' = \boldsymbol{S}_{11} \cdot \boldsymbol{T}^{\mathrm{T}} \cdot \boldsymbol{d}_1' + \boldsymbol{S}_{12} \cdot \boldsymbol{T}^{\mathrm{T}} \cdot \boldsymbol{d}_2'$$
$$\boldsymbol{T}^{\mathrm{T}} \cdot \boldsymbol{x}_2' = \boldsymbol{S}_{21} \cdot \boldsymbol{T}^{\mathrm{T}} \cdot \boldsymbol{d}_1' + \boldsymbol{S}_{22} \cdot \boldsymbol{T}^{\mathrm{T}} \cdot \boldsymbol{d}_2'$$

multiplying both sides by $\boldsymbol{T}$ and putting $\boldsymbol{S}' = \boldsymbol{T} \cdot \boldsymbol{S} \cdot \boldsymbol{T}^{\mathrm{T}}$ we have

$$\boldsymbol{x}_1' = \boldsymbol{S}_{11}' \cdot \boldsymbol{d}_1' + \boldsymbol{S}_{12}' \cdot \boldsymbol{d}_2'$$
$$\boldsymbol{x}_2' = \boldsymbol{S}_{21}' \cdot \boldsymbol{d}_1' + \boldsymbol{S}_{22}' \cdot \boldsymbol{d}_2' \tag{27}$$

The matrix S'_{21} will now have the form

$$\begin{bmatrix} \frac{2EI}{L} & C\cdot\frac{6EI}{L^2} & -S\cdot\frac{6EI}{L^2} \\ -C\cdot\frac{6EI}{L^2} & -C^2\cdot\frac{12EI}{L^3}-S^2\cdot\frac{EA}{L} & S\cdot C\cdot\left(\frac{12EI}{L^3}-\frac{EA}{L}\right) \\ S\cdot\frac{6EI}{L^2} & S\cdot C\cdot\left(\frac{12EI}{L^3}-\frac{EA}{L}\right) & -C^2\cdot\frac{EA}{L}-S^2\cdot\frac{12EI}{L^3} \end{bmatrix}$$

where $C=\text{Cos }\alpha$ and $S=\text{Sin }\alpha$
The matrices S'_{11}, S'_{12}, and S'_{22} are similar. For the frame shown in Fig. 37 the appropriate values of α are

member a	$\alpha=90°$
member b	$\alpha=90°$
member c	$\alpha=0$
member d	$\alpha=90°$
member e	$\alpha=90°$
member f	$\alpha=\tan^{-1}$ (length of a/length of g)
member g	$\alpha=0$

It is clear that where the member co-ordinates are at right angles to the structure co-ordinates, the transformation of the stiffness matrices, $\mathbf{S}_{11}$, etc., is simple. A high proportion of structures contain only members at right angles and therefore this step does not involve a large amount of calculation.

Having obtained the stiffness matrices for each member we must build up the stiffness matrix ($\mathbf{S}$) for the complete structure. Each joint is rigid and hence the displacements at the ends of the members must equal the displacements of the appropriate joints.

Applying this, the condition of compatibility, the relationships between the member end displacements and joint displacements are

member a	$\mathbf{d}'_1=\mathbf{o}$	$\mathbf{d}'_2=\mathbf{d}_A$	
member b	$\mathbf{d}'_1=\mathbf{d}_A$	$\mathbf{d}'_2=\mathbf{d}_B$	
member c	$\mathbf{d}'_1=\mathbf{d}_B$	$\mathbf{d}'_2=\mathbf{d}_C$	
member d	$\mathbf{d}'_1=\mathbf{d}_D$	$\mathbf{d}'_2=\mathbf{d}_C$	(28)
member e	$\mathbf{d}'_1=\mathbf{o}$	$\mathbf{d}'_2=\mathbf{d}_D$	
member f	$\mathbf{d}'_1=\mathbf{o}$	$\mathbf{d}'_2=\mathbf{d}_D$	
member g	$\mathbf{d}'_1=\mathbf{d}_A$	$\mathbf{d}'_2=\mathbf{d}_D$	

Substituting from these equations for the member end-displacements in equation (27) gives

$$\begin{array}{ll}
\text{member } a & \boldsymbol{x}'_{2a}=(S'_{22})_a\boldsymbol{d}_A \\
\text{member } b & \boldsymbol{x}'_{1b}=(S'_{11})_b\boldsymbol{d}_A+(S'_{12})_b\boldsymbol{d}_B \\
 & \boldsymbol{x}'_{2b}=(S'_{21})_b\boldsymbol{d}_A+(S'_{22})_b\boldsymbol{d}_B \\
\text{member } c & \boldsymbol{x}'_{1c}=(S'_{11})_c\boldsymbol{d}_B+(S'_{12})_c\boldsymbol{d}_C \\
 & \boldsymbol{x}'_{2c}=(S'_{21})_c\boldsymbol{d}_B+(S'_{22})_c\boldsymbol{d}_C \\
\text{member } d & \boldsymbol{x}'_{1d}=(S'_{11})_d\boldsymbol{d}_D+(S'_{12})_d\boldsymbol{d}_C \\
 & \boldsymbol{x}'_{2d}=(S'_{21})_d\boldsymbol{d}_D+(S'_{22})_d\boldsymbol{d}_C \\
\text{member } e & \boldsymbol{x}'_{2e}=(S'_{22})_e\boldsymbol{d}_D \\
\text{member } f & \boldsymbol{x}'_{2f}=(S'_{22})_f\boldsymbol{d}_D \\
\text{member } g & \boldsymbol{x}'_{1g}=(S'_{11})_g\boldsymbol{d}_A+(S'_{12})_g\boldsymbol{d}_D \\
 & \boldsymbol{x}'_{2g}=(S'_{21})_g\boldsymbol{d}_A+(S'_{22})_g\boldsymbol{d}_D
\end{array} \tag{29}$$

The primes have been omitted from the joint displacements $\boldsymbol{d}_A$ etc., since these are obviously in structure co-ordinates. The expressions for $\boldsymbol{x}'_1$ in members a, e and f have been omitted since they are foundation reactions and do not enter into the joint equilibrium equations.

At each joint the sum of the member end-loads at that joint must equal the external force applied to the joint. Applying this condition, the joint equilibrium equations are

$$\begin{array}{ll}
\boldsymbol{x}'_{2a}+\boldsymbol{x}'_{1b}+\boldsymbol{x}'_{1g}=\boldsymbol{P}_A & \boldsymbol{x}'_{2c}+\boldsymbol{x}'_{2d}=\boldsymbol{P}_C \\
\boldsymbol{x}'_{2b}+\boldsymbol{x}'_{1c}=\boldsymbol{P}_B & \boldsymbol{x}'_{1d}+\boldsymbol{x}'_{2e}+\boldsymbol{x}'_{2f}+\boldsymbol{x}'_{2g}=\boldsymbol{o}
\end{array} \tag{30}$$

Again the primes are omitted from $\boldsymbol{P}_A$, $\boldsymbol{P}_B$ and $\boldsymbol{P}_C$.
Substituting from equations (29) into these expressions and writing the resulting equations in matrix form gives

$$\begin{bmatrix}\boldsymbol{P}_A\\ \boldsymbol{P}_B\\ \boldsymbol{P}_C\\ \boldsymbol{o}\end{bmatrix}=$$

$$\begin{bmatrix}
(S'_{22})_a+(S'_{11})_b+(S'_{11})_g & (S'_{12})_b & 0 & (S'_{12})_g \\
(S'_{21})_b & (S'_{22})_b+(S'_{11})_c & (S'_{12})_c & 0 \\
0 & (S'_{21})_c & (S'_{22})_c+(S'_{22})_d & (S'_{21})_d \\
(S'_{21})_g & 0 & (S'_{12})_d & (S'_{11})_d+(S'_{22})_e+(S'_{22})_f+(S'_{22})_g
\end{bmatrix}\times$$

$$\begin{bmatrix}\boldsymbol{d}_A\\ \boldsymbol{d}_B\\ \boldsymbol{d}_C\\ \boldsymbol{d}_D\end{bmatrix} \qquad \text{i.e.}\quad \boldsymbol{P}=\boldsymbol{S}\cdot\boldsymbol{d} \tag{31}$$

The form of the stiffness matrix for the complete structure reflects the arrangement of the members within the structure. The terms on the leading diagonal correspond to the joints and those off the leading diagonal correspond to the connections between joints. Members a, b and g contribute to the stiffness of joint A and thus their $\boldsymbol{S}'$ matrices appear in the first leading diagonal term. Furthermore only the matrix $\boldsymbol{S}'$ corresponding to the end of the member at that joint is present. By comparing Fig. 37 with equations (31) it is clear that the same relationship exists for all the leading diagonal terms. Joints A and D are connected by member g. Correspondingly the $\boldsymbol{S}'$ matrix for member g appears in the row and column common to both A and D. It will be noticed that the off-diagonal terms only contain a single $\boldsymbol{S}'$ matrix signifying that there is only one member connecting any two joints or indeed no connection as the case may be. The subscripts of the off-diagonal $\boldsymbol{S}'$ matrices are all unequal. This indicates that they represent the loads at one end of a member set up by displacements at the other end, as would be expected.

It will be noted that the matrix $\boldsymbol{S}$ is symmetric, this follows from the reciprocal theorem.

The relationship between the stiffness matrix and the geometry of the structure means that the complete matrix may be written down in terms of member stiffness matrices simply by inspection. The necessity of formally setting down the conditions of compatibility (equations 29) and equilibrium (equations 30) is therefore avoided.

All that remains to be done to find the loads in each member is to solve equations (31). This may be done directly, or by inverting the stiffness matrix. Where a structure is to be analysed for several different loading cases, it is usually better to invert the matrix $\boldsymbol{S}$. Having obtained the inverse, the displacements resulting from each loading are found from the equation.

$$\boldsymbol{d}=\boldsymbol{S}^{-1}\cdot\boldsymbol{P}$$

Knowing $\boldsymbol{d}$ the stress resultants in each member can be found from equations (29).

Summary of procedure for the stiffness method

Basic Outline

1. Determine α_k.
2. Mark on each member an arrow and a letter by which it may be identified.

3. Calculate the relationship between the end loads and the end displacements for each member

$$\boldsymbol{x}_1 = \boldsymbol{S}_{11}\boldsymbol{d}_1 + \boldsymbol{S}_{12}\boldsymbol{d}_2$$
$$\boldsymbol{x}_2 = \boldsymbol{S}_{21}\boldsymbol{d}_1 + \boldsymbol{S}_{22}\boldsymbol{d}_2$$

4. Transform these equations from member to structure co-ordinates where necessary.
5. Apply the conditions of compatibility.
6. Set up the equations of joint equilibrium.

$$\boldsymbol{P} = \boldsymbol{S} \cdot \boldsymbol{d}$$

7. Invert the stiffness matrix $\boldsymbol{S}$.
8. Calculate the member end displacements.

$$\boldsymbol{d} = \boldsymbol{S}^{-1} \cdot \boldsymbol{P}$$

9. Determine the stress resultants in each member from the equations developed in stage 4.

Detailed Procedure

1. It is not as easy to identify the constraints in a structure as it is to pick out the releases. The best way to calculate the kinematical indeterminacy is to add up the number of degrees of freedom at each node. Remember that where a load is applied there must be a node, and that the ground node has no degrees of freedom.

2. This stage is self-explanatory, but it makes the calculation easier to follow if each member has a unique reference. For each member, end 1 is at the tail of the arrow and end 2 is at the head.

3. Obtain the equations relating the end-loads in each member to the end-displacements in terms of the member co-ordinates.

$$\boldsymbol{x}_1 = \boldsymbol{S}_{11}\boldsymbol{d}_1 + \boldsymbol{S}_{12}\boldsymbol{d}_2$$
$$\boldsymbol{x}_2 = \boldsymbol{S}_{21}\boldsymbol{d}_1 + \boldsymbol{S}_{22}\boldsymbol{d}_2$$

There are as many terms in $\boldsymbol{d}_1$ and $\boldsymbol{d}_2$ as there are degrees of freedom at each node.

4. Transform the relationships developed in stage 3 above into structure co-ordinates. This is carried out by applying the transformation

$$\boldsymbol{S}' = \boldsymbol{T} \cdot \boldsymbol{S} \cdot \boldsymbol{T}^{\mathrm{T}}$$

where

$$\boldsymbol{T} = \begin{bmatrix} 1 & 0 & 0 \\ 0 & \text{Cos}\ \alpha & \text{Sin}\ \alpha \\ 0 & -\text{Sin}\ \alpha & \text{Cos}\ \alpha \end{bmatrix}$$

α = the angle of inclination of the member with respect to the positive horizontal axis of the structure measured positive anti-clockwise.

5. The condition of compatibility is that the end displacements of a member must be the same as the displacements of the nodes at the ends of the member. In effect this simply means changing the notation in the expressions obtained above, as shown in the following example

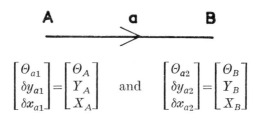

$$\begin{bmatrix} \Theta_{a1} \\ \delta y_{a1} \\ \delta x_{a1} \end{bmatrix} = \begin{bmatrix} \Theta_A \\ Y_A \\ X_A \end{bmatrix} \quad \text{and} \quad \begin{bmatrix} \Theta_{a2} \\ \delta y_{a2} \\ \delta x_{a2} \end{bmatrix} = \begin{bmatrix} \Theta_B \\ Y_B \\ X_B \end{bmatrix}$$

6. The end loads in the members meeting at a node must be in equilibrium with the external loads acting on that node.

$$\boldsymbol{P} = \sum \boldsymbol{x}$$

Having set down the equation for equilibrium at each node, by substituting for the end loads from expressions obtained in stage 5 and combining the equations, the relationship between the applied loads and the displacements of the structure is obtained. In matrix form, this is the equation

$$\boldsymbol{P} = \boldsymbol{S} \cdot \boldsymbol{d}$$

7. Invert the stiffness matrix, $\boldsymbol{S}$, either by one of the methods given in Part I or by computer, depending upon the order of the matrix.

8. The values of the member end displacements are found from the equation

$$\boldsymbol{d} = \boldsymbol{S}^{-1} \cdot \boldsymbol{P}$$

The stiffness matrix is independent of the loading and, like the flexibility matrix, depends only on the geometry. Thus, once the inverse of the stiffness matrix has been calculated, the structure may be analysed for any other loading quite readily.

9. The stress resultants in each member may now be obtained by substituting the values for $\boldsymbol{d}$ into the equations

$$\boldsymbol{x}_1' = \boldsymbol{S}_{11}' \boldsymbol{d}_1' + \boldsymbol{S}_{12} \boldsymbol{d}_2' \text{ etc.}$$

The primes indicate the equation is related to the structure co-ordinates. The displacements calculated in stage 8 are in structure co-ordinates and, therefore, the equations obtained in stage 4 and not those in stage 3 must be used.

Worked Example C

This example is repeated using the stiffness method in order to give a direct comparison. The complexity of this solution shows clearly that this is the wrong method for this example. The solution was obtained, using a desk calculating machine, to six significant figures after the decimal. This order of accuracy is essential in order to obtain an accurate answer. For convenience and simplicity, only four significant figures after the decimal are given here.

Solution

There are two nodes, apart from the ground node, which each have three degrees of freedom, as shown in Fig. 40. The kinematical indeterminacy therefore equals six.

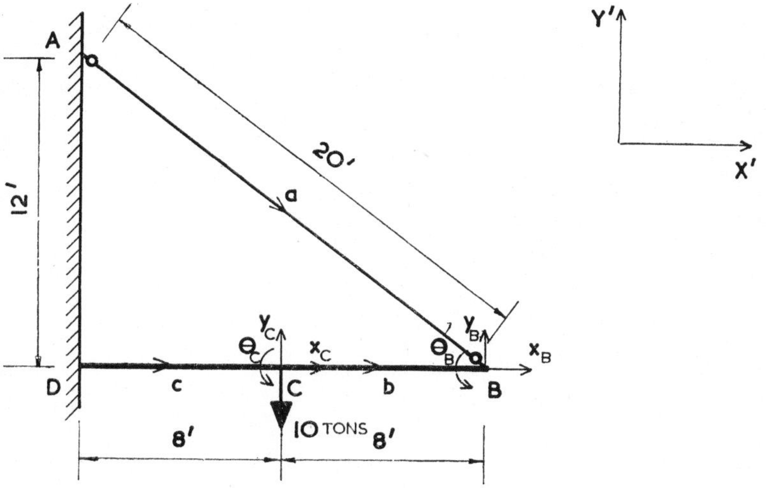

FIG. 40

Using the same member properties as before, i.e.

Moment of Inertia of beam $= 300$ ins^4
Cross sectional area of beam $=$ 10 ins^2
Cross sectional area of tie $=$ 1·0 ins^2

the equations, $\boldsymbol{x}_1 = \boldsymbol{S}_{11}\boldsymbol{d}_1 + \boldsymbol{S}_{12}\boldsymbol{d}_2$ etc. become

$$\begin{bmatrix} n_{c2} \\ s_{c2} \\ m_{c2} \end{bmatrix} = E \begin{bmatrix} \frac{10}{12\times 8} & 0 & 0 \\ 0 & \frac{12\times 300}{12^3\times 8^3} & -\frac{6\times 300}{12^2\times 8^2} \\ 0 & -\frac{6\times 300}{12^2\times 8^2} & \frac{4\times 300}{12\times 8} \end{bmatrix} \begin{bmatrix} x_{c2} \\ y_{c2} \\ \Theta_{c2} \end{bmatrix}$$

Note that since end 1 of member c is attached to the "ground" the stress resultants there are ground reactions and they do not influence the analysis.

$$\begin{bmatrix} n_{b1} \\ s_{b1} \\ m_{b1} \end{bmatrix} = E \begin{bmatrix} \frac{10}{12\times 8} & 0 & 0 \\ 0 & \frac{12\times 300}{12^3\times 8^3} & \frac{6\times 300}{12^2\times 8^2} \\ 0 & \frac{6\times 300}{12^2\times 8^2} & \frac{4\times 300}{12\times 8} \end{bmatrix} \begin{bmatrix} x_{b1} \\ y_{b1} \\ \Theta_{b1} \end{bmatrix}$$

$$+E \begin{bmatrix} -\frac{10}{12\times 8} & 0 & 0 \\ 0 & -\frac{12\times 300}{12^3\times 8^3} & \frac{6\times 300}{12^2\times 8^2} \\ 0 & -\frac{6\times 300}{12^2\times 8^2} & \frac{2\times 300}{12\times 8} \end{bmatrix} \begin{bmatrix} x_{b2} \\ y_{b2} \\ \Theta_{b2} \end{bmatrix}$$

$$\begin{bmatrix} n_{b2} \\ s_{b2} \\ m_{b2} \end{bmatrix} = E \begin{bmatrix} -\frac{10}{12\times 8} & 0 & 0 \\ 0 & -\frac{12\times 300}{12^3\times 8^3} & -\frac{6\times 300}{12^2\times 8^2} \\ 0 & \frac{6\times 300}{12^2\times 8^2} & \frac{2\times 300}{12\times 8} \end{bmatrix} \begin{bmatrix} x_{b1} \\ y_{b1} \\ \Theta_{b1} \end{bmatrix}$$

$$+E \begin{bmatrix} \frac{10}{12\times 8} & 0 & 0 \\ 0 & \frac{12\times 300}{12^3\times 8^3} & -\frac{6\times 300}{12^2\times 8^2} \\ 0 & -\frac{6\times 300}{12^2\times 8^2} & \frac{4\times 300}{12\times 8} \end{bmatrix} \begin{bmatrix} x_{b2} \\ y_{b2} \\ \Theta_{b2} \end{bmatrix}$$

$$\begin{bmatrix} n_{a2} \\ s_{a2} \\ m_{a2} \end{bmatrix} = E \begin{bmatrix} \frac{1}{12\times 20} & 0 & 0 \\ 0 & 0 & 0 \\ 0 & 0 & 0 \end{bmatrix} \begin{bmatrix} x_{a2} \\ y_{a2} \\ \Theta_{a2} \end{bmatrix}$$

The only equation which requires transformation into structure co-ordinates is that for member a. The angle between the X axis and the member is $-\alpha$, where $\alpha = \tan^{-1} 0{\cdot}75$.

Therefore, the transformation matrix is

$$\boldsymbol{T} = \begin{bmatrix} \cos\alpha & -\sin\alpha & 0 \\ \sin\alpha & \cos\alpha & 0 \\ 0 & 0 & 1 \end{bmatrix} = \begin{bmatrix} 0{\cdot}8 & 0{\cdot}6 & 0 \\ -0{\cdot}6 & 0{\cdot}8 & 0 \\ 0 & 0 & 1 \end{bmatrix}$$

The equation for member a becomes

$$\begin{bmatrix} n'_{a2} \\ s'_{a2} \\ m'_{a2} \end{bmatrix} = E \begin{bmatrix} \frac{0\cdot 64}{240} & -\frac{0\cdot 48}{240} & 0 \\ -\frac{0\cdot 48}{240} & \frac{0\cdot 36}{240} & 0 \\ 0 & 0 & 0 \end{bmatrix} \begin{bmatrix} x'_{a2} \\ y'_{a2} \\ \Theta'_{a2} \end{bmatrix}$$

All that need be done to apply the conditions of compatibility is to change the notation.

For instance

$$\begin{bmatrix} x'_{b2} \\ y'_{b2} \\ \Theta'_{b2} \end{bmatrix} \quad \text{becomes} \quad \begin{bmatrix} X_B \\ Y_B \\ \Theta_B \end{bmatrix}$$

The equations of joint equilibrium are

$$\begin{aligned} n'_{c2} + n'_{b1} &= 0 \\ s'_{c2} + s'_{b1} &= -10 \\ m'_{c2} + m'_{b1} &= 0 \\ n'_{a2} + n'_{b2} &= 0 \\ s'_{a2} + s'_{b2} &= 0 \\ m'_{a2} + m'_{b2} &= 0 \end{aligned}$$

The primes indicate that all the terms are in structure co-ordinates.

Combining these equations and substituting for the terms on the left hand side gives, in matrix form

$$\boldsymbol{P} = \boldsymbol{S} \cdot \boldsymbol{d}$$

$$\begin{bmatrix} 0 \\ -10 \\ 0 \\ 0 \\ 0 \\ 0 \end{bmatrix} = E \begin{bmatrix} \frac{5}{24} & 0 & 0 & -\frac{5}{48} & 0 & 0 \\ 0 & \frac{25}{3072} & 0 & 0 & -\frac{25}{6144} & \frac{25}{128} \\ 0 & 0 & 25 & 0 & -\frac{25}{128} & \frac{25}{4} \\ -\frac{5}{48} & 0 & 0 & \left(\frac{1}{375}+\frac{5}{48}\right) & -\frac{1}{500} & 0 \\ 0 & -\frac{25}{6144} & -\frac{25}{128} & -\frac{1}{500} & \left(\frac{3}{2000}+\frac{25}{6144}\right) & -\frac{25}{128} \\ 0 & \frac{25}{128} & \frac{25}{4} & 0 & -\frac{25}{128} & \frac{25}{2} \end{bmatrix} \begin{bmatrix} X_C \\ Y_C \\ \Theta_C \\ X_B \\ Y_B \\ \Theta_B \end{bmatrix}$$

Inverting $\boldsymbol{S}$ and multiplying this equation through by $\boldsymbol{S}^{-1}$ we obtain

$$\boldsymbol{d} = \boldsymbol{S}^{-1} \cdot \boldsymbol{P}$$

$$\begin{bmatrix} X_C \\ Y_C \\ \Theta_C \\ X_B \\ Y_B \\ \Theta_B \end{bmatrix} =$$

$$\frac{1}{E}\begin{bmatrix} 9{\cdot}5809 & 3{\cdot}6725 & 0{\cdot}06886 & 9{\cdot}5618 & 11{\cdot}7523 & 0{\cdot}09182 \\ 3{\cdot}6725 & 277{\cdot}8759 & 2{\cdot}1382 & 7{\cdot}3449 & 201{\cdot}0768 & -2{\cdot}2691 \\ 0{\cdot}06886 & 2{\cdot}1382 & 0{\cdot}07157 & 0{\cdot}1377 & 3{\cdot}7704 & -0{\cdot}01054 \\ 9{\cdot}5618 & 7{\cdot}3449 & 0{\cdot}1377 & 19{\cdot}1234 & 23{\cdot}5045 & 0{\cdot}1836 \\ 11{\cdot}7523 & 201{\cdot}0768 & 3{\cdot}7704 & 23{\cdot}5045 & 643{\cdot}4719 & 5{\cdot}0273 \\ 0{\cdot}09182 & -2{\cdot}2691 & -0{\cdot}01054 & 0{\cdot}1836 & 5{\cdot}0273 & 0{\cdot}1993 \end{bmatrix} \begin{bmatrix} 0 \\ -10 \\ 0 \\ 0 \\ 0 \\ 0 \end{bmatrix}$$

hence

$$\begin{bmatrix} X_C \\ Y_C \\ \Theta_C \\ X_B \\ Y_B \\ \Theta_B \end{bmatrix} = \frac{1}{E}\begin{bmatrix} -\ \ 36{\cdot}7245 \\ -2778{\cdot}7587 \\ -\ \ 21{\cdot}3822 \\ -\ \ 73{\cdot}4486 \\ -2010{\cdot}7677 \\ 22{\cdot}6911 \end{bmatrix}$$

Substituting the values of the displacements into the equations for each member gives

$$\begin{aligned} n_{c2} &= -\ 3{\cdot}826 \\ s_{c2} &= -\ 7{\cdot}131 \\ m_{c2} &= 22{\cdot}954 \\ n_{b1} &= 3{\cdot}826 \\ s_{b1} &= 2{\cdot}869 \\ m_{b1} &= 22{\cdot}954 \\ n_{b2} &= -\ 3{\cdot}826 \\ s_{b2} &= 2{\cdot}869 \\ m_{b2} &= 0 \\ n_{a2} &= 4{\cdot}782 \\ s_{a2} &= 0 \\ m_{a2} &= 0 \end{aligned}$$

These values closely agree with those obtained by the Flexibility approach.

There is no check for this method corresponding to that for the flexibility method. However, for this example, a similar check to that used previously could be applied.

SOLUTION TO EXERCISES SHOWN IN FIG. 15.

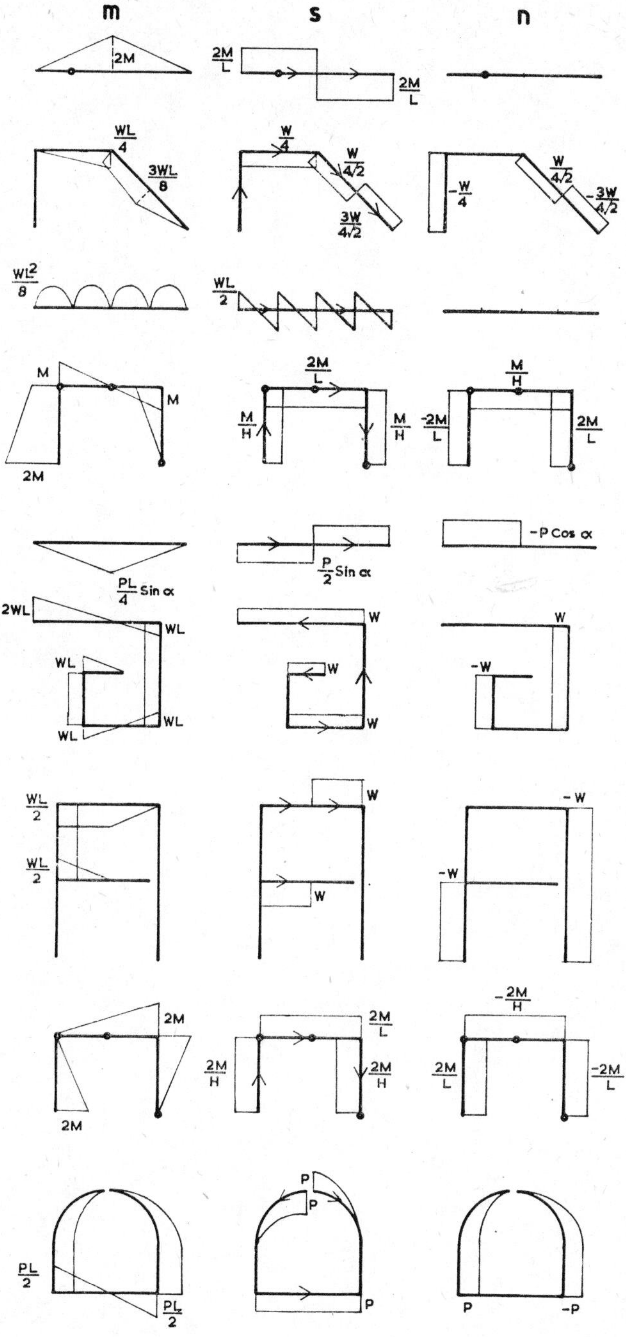

BIBLIOGRAPHY

1. LIVESLEY, R. K., *Matrix Methods of Structural Analysis,* Pergamon Press, 1964.

2. MORICE, P. B., *Linear Structural Analysis,* Thames & Hudson, London, 1959.

3. HALL, A. S. and WOODHEAD, R. W., *Frame Analysis,* Wiley, New York.

4. McMINN, S. J., *Matrices for Structural Analysis,* Spon, London, 1962.

5. HENDERSON, J. C. de C. and BICKLEY, W. G. 'Statical Indeterminacy of a Structure'. *Aircraft Engineering,* December 1955.

6. AITKEN, A. C., *Determinants and Matrices,* 9th. ed., OLIVER and BOYD, EDINBURGH, 1958.

7. MUNRO, J. 'The Elastic and Limit Analysis of Planar Skeletal Structures'. *Civil Engineering,* May 1965.

8. JENKINS, R. S., Taylor Woodrow Foundation Lectures, Nottingham University, 1961. 'Matrix Methods in Structural Mechanics'.